There is not any black hole

CARLO MARIA PACE

THERE IS NOT
ANY BLACK HOLE

Title | There is not any black hole
Author | Carlo Maria Pace

ISBN | 978-88-92642-24-9

Youcanprint Self-Publishing
Via Roma, 73 – 73039 Tricase (LE) – Italy
www.youcanprint.it
info@youcanprint.it
Facebook: facebook.com/youcanprint.it
Twitter: twitter.com/youcanprintit

INTRODUCTION

In this treatment I, by starting from the Einstein's equations (or more precisely from the Einstein's field equation) of the General Theory of Relativity[1], demonstrate that the so-called Schwarzschild solution[2] is incorrect by means of calculating the correct solution for the case that Schwarzschild has analysed[3]. In particular, I demonstrate that, in the correct Schwarzschild solution, there is not any event horizon and therefore there is not any black hole[4]. Moreover, I extend all this to its direct consequences, in particular correcting the Kerr solution[5], the Reissner-Nordstrøm solution[6] and the Kerr-Newman solution[7]. Then, I extend these four correct solutions (of Schwarzschild, of Kerr, of Reissner-Nordstrøm and of Kerr-Newman) also to the case in which the cosmological constant is greater than zero[8].

On the other hand, I, by starting from the general form of the Einstein's field equation of the General Theory of Relativity[9],

---

[1] Cf. EINSTEIN (1916); OHANIAN – RUFFINI pp. 275-293; MISNER – THORNE – WHEELER pp. 383-590; WEINBERG (1972) pp. 151-171; ANDERSON pp. 329-371.

[2] Cf. SCHWARZSCHILD; OHANIAN – RUFFINI pp. 293-298; MISNER – THORNE – WHEELER pp. 636-687.819-840; WEINBERG (1972) pp. 175-182; ANDERSON pp. 381-392.

[3] Cf. SCHWARZSCHILD.

[4] Cf. OHANIAN – RUFFINI pp. 324-342; MISNER – THORNE – WHEELER pp. 842-940; WEINBERG (1972) pp. 207-208.297.349.479.

[5] Cf. OHANIAN – RUFFINI pp. 343-346.349-360; MISNER – THORNE – WHEELER pp. 877-878; WEINBERG (1972) pp. 240.349.

[6] Cf. OHANIAN – RUFFINI pp. 343.346-348.360; MISNER – THORNE – WHEELER pp. 840-841.920-921.

[7] Cf. MISNER – THORNE – WHEELER pp. 875-885.891-894; OHANIAN – RUFFINI pp. 348.355.

[8] Cf. OHANIAN – RUFFINI p. 299; WEINBERG (1972) pp. 613-616.

[9] Cf. EINSTEIN (1916); OHANIAN – RUFFINI pp. 275-293; MISNER – THORNE – WHEELER pp. 383-590; WEINBERG (1972) pp. 151-171; ANDERSON pp. 329-371.

demonstrate also in general the non-existence of any event horizon, and therefore the non-existence of any black hole, in the field of application of the General Theory of Relativity.

Finally, I also explore the consequences of all this on the entropy of the Universe[10], on the Hawking emission process[11], on the Big Bang Theory[12] and on the astronomical observations.

---

[10] Cf. OHANIAN – RUFFINI pp. 362-363.422; HAWKING – PENROSE pp. 32-36.49-74.105.114-116; BARROW (1998) pp. 31-39; DAVIES (2000) p. 149.

[11] Cf. OHANIAN – RUFFINI pp. 360-367; HAWKING – PENROSE pp. 49-74; HAWKING (1997) pp. 120-135.

[12] Cf. OHANIAN – RUFFINI pp. 389-437.444-473; MISNER – THORNE – WHEELER pp. 703-799; WEINBERG (1972) pp. 469-597; ANDERSON pp. 444-465; HAWKING – PENROSE pp. 30.37-47.91-122; BARROW (1998) pp. 11-125; DAVIES (2000) pp. 30-46; HAWKING (1997) pp. 51-70.136-165; WEINBERG (1980) pp. 14-165; DAVIES (1997) pp. 133-177.200-214; GOTT pp. 153-197.

CHAPTER I
THE CORRECT SCHWARZSCHILD SOLUTION OF THE EINSTEIN'S FIELD EQUATION

1 The correct Schwarzschild solution

The Schwarzschild solution (of the Einstein's field equation of the General Theory of Relativity) is the solution (of the Einstein's field equation of the General Theory of Relativity) for the gravitational field surrounding a spherically symmetric mass distribution[13].

Therefore, for obtaining the correct Schwarzschild solution, in what follows we will solve the Einstein's equations, or more precisely the Einstein's field equation, of the General Theory of Relativity in the exterior of a spherically symmetric mass distribution, that is, in the vacuum region surrounding the spherically symmetric mass distribution.

Now, if the mass distribution is spherically symmetric, the gravitational field generated by this mass distribution must also have spherical symmetry. Since it is sometimes of interest to deal with a collapsing spherical mass – that is, a sphere that shrinks in size – we will begin with an *Ansatz* (that is, with an approach) for

---

[13] For calculating the correct Schwarzschild solution (of the Einstein's field equation of the General Theory of Relativity) I will correct the treatment that has been made by H. C. Ohanian and R. Ruffini for obtaining the Schwarzschild solution (of the Einstein's field equation of the General Theory of Relativity): cf. OHANIAN – RUFFINI pp. 293-298.

the space-time metric that includes a time dependence for the components of the metric tensor. However, at the end of our calculation we will find that in the exterior region this time dependence disappears, because, as in the case of the Newtonian gravitational potential, the exterior solution for the metric tensor of a spherical mass does not depend on the size of the mass and remains static even when the mass collapses.

The spherical symmetry of the gravitational field imposes severe restrictions on the form of the space-time interval. The rectangular displacements dx, dy, dz must occur in the combination $dx^2 + dy^2 + dz^2$ characteristic of spherical symmetry; this combination takes the familiar form $dr^2 + r^2 d\theta^2 + r^2 \sin^2\theta\, d\varphi^2$ when is expressed in spherical polar coordinates. Furthermore, although the product $drdt$ is consistent with spherical symmetry, the products $d\theta dt$ and $d\varphi dt$ are not consistent with spherical symmetry, because they would imply that a light signal (with $ds^2 = 0$) has different speeds when moving in the direction of increasing or decreasing θ and φ. Accordingly, the space-time interval must be of the form:

$$ds^2 = A(r,t)dt^2 - B(r,t)[dr^2 + r^2 d\theta^2 +$$

$$+ r^2 \sin^2\theta\, d\varphi^2] - 2F(r,t)drdt \tag{1}$$

where $A(r,t)$, $B(r,t)$ and $F(r,t)$ are some functions of the radial coordinate r and the time coordinate t.

The angular coordinates θ and φ are unambiguous; the measurement of these coordinates depends only on our ability to divide a circumference concentric with the mass into equal parts, which we can do even when the functions $A(r,t)$, $B(r,t)$ and $F(r,t)$ are not specified. The radial coordinate r is ambiguous because we do not yet know its precise relation to the measurement of distance. For a start, we will treat r simply as a parameter that identifies different spherical surfaces concentric with the mass. The time coordinate t suffers from similar ambiguities. However, we will insist that when $r \to +\infty$ and the space-time becomes flat, the increments in r and t should equal the increments in the true distance and time, respectively. This requires that for $r \to +\infty$, $A(r,t)$ and $B(r,t) \to 1$.

The term involving $drdt$ in the equation (1) can be eliminated by a change in the time coordinate. We introduce a new time coordinate $\tilde{t}$ such that:

$$d\tilde{t} = (Adt - Fdr)Q(r,t) \tag{2}$$

where $Q(r,t)$ is a function of r and t that is to be chosen so as to make the right side of the equation (2) into a perfect differential [this requires that Q satisfy the following differential equation: $(\partial/\partial t)(FQ) = -(\partial/\partial r)(AQ)$].

From the equation (2) we obtain:

$$Adt^2 - 2Fdrdt = \frac{1}{Q^2 A}d\tilde{t}^2 - \frac{F^2}{A}dr^2 \tag{3}$$

which gives:

$$ds^2 = \frac{1}{Q^2 A}d\tilde{t}^2 - \left(B + \frac{F^2}{A}\right)dr^2 - B[r^2 d\theta^2 +$$

$$+ r^2 \sin^2\theta \, d\varphi^2] \tag{4}$$

We can simplify the equation (4) further by introducing a new radial coordinate $\tilde{r} = r\sqrt{B(r)}$. This has the advantage that the terms involving angular displacements reduce to $\tilde{r}^2 d\theta^2 + \tilde{r}^2 \sin^2\theta \, d\varphi^2$, so we obtain:

$$ds^2 = \frac{1}{Q^2 A} d\tilde{t}^2 - \left(B + \frac{F^2}{A}\right)\left(\frac{\partial r}{\partial \tilde{r}}\right)^2 d\tilde{r}^2 - \tilde{r}^2 d\theta^2 +$$

$$- \tilde{r}^2 \sin^2 \theta \, d\varphi^2 \qquad (5)$$

For the solution of the Einstein's equations, or more precisely of the Einstein's field equation, of the General Theory of Relativity, it is convenient to omit the tildes in the equation (5) and to write the unknown functions multiplying dt^2 and dr^2 as exponentials:

$$ds^2 = e^{N(r,t)} dt^2 - e^{L(r,t)} dr^2 - r^2 d\theta^2 +$$

$$- r^2 \sin^2 \theta \, d\varphi^2 \qquad (6)$$

With $x^0 = t$, $x^1 = r$, $x^2 = \theta$, $x^3 = \varphi$, the metric tensor corresponding to the equation (6) is:

$$g_{\mu\nu} = \begin{pmatrix} e^N & 0 & 0 & 0 \\ 0 & -e^L & 0 & 0 \\ 0 & 0 & -r^2 & 0 \\ 0 & 0 & 0 & -r^2 \sin^2 \theta \end{pmatrix} \qquad (7)$$

The unknown functions are now $N(r,t)$ and $L(r,t)$. We will use the Einstein's equations, or more precisely the Einstein's field equation, of the General Theory of Relativity to find them.

The Christoffel symbols for a metric tensor of the form (7) are (the primes indicate derivatives, $N' \equiv \dfrac{\partial N}{\partial r}$, $L' \equiv \dfrac{\partial L}{\partial r}$; the dots indicate time derivatives, $\dot{N} \equiv \dfrac{\partial N}{\partial t}$, $\dot{L} \equiv \dfrac{\partial L}{\partial t}$):

$$\Gamma^0{}_{00} = \tfrac{1}{2}\,\dot{N}\,, \quad \Gamma^0{}_{01} = \Gamma^0{}_{10} = \tfrac{1}{2}\,N'\,,$$

$$\Gamma^0{}_{11} = \tfrac{1}{2}\,\dot{L}\,e^{L-N}\,, \quad \Gamma^1{}_{00} = \tfrac{1}{2}\,N'e^{N-L}\,,$$

$$\Gamma^1{}_{10} = \Gamma^1{}_{01} = \tfrac{1}{2}\,\dot{L}\,, \quad \Gamma^1{}_{11} = \tfrac{1}{2}\,L'\,,$$

$$\Gamma^1{}_{22} = -\,r\,e^{-L}\,, \quad \Gamma^1{}_{33} = -\,r\sin^2\theta\,e^{-L} \qquad (8)$$

$$\Gamma^2{}_{12} = \Gamma^2{}_{21} = \tfrac{1}{r}\,, \quad \Gamma^2{}_{33} = -\,\sin\theta\cos\theta$$

$$\Gamma^3{}_{13} = \Gamma^3{}_{31} = \tfrac{1}{r}\,, \quad \Gamma^3{}_{23} = \Gamma^3{}_{32} = \cot\theta$$

All other Christoffel symbols are zero. The Ricci tensor $R_{\mu\nu}$ can be calculated from the Christoffel symbols; only the 00, 01, 10, 11, 22, 33 components of $R_{\mu\nu}$ are nonzero.

The Einstein's field equation (of the General Theory of Relativity) in vacuum is:

$$R_{\mu\nu} - \tfrac{1}{2}\,g_{\mu\nu}\,R = 0 \qquad (9)$$

By taking the trace of this equation, we find that $R = 0$. Hence the equation (9) reduces to:

$$R_{\mu\nu} = 0 \qquad (10)$$

The 00, 01, 10, 11, 22 and 33 components of this equation are as follows:

$$R_{00} = e^{N-L}\left(-\frac{N''}{2} + \frac{L'N'}{4} - \frac{N'^2}{4} - \frac{N'}{r}\right) + \frac{\ddot{L}}{2} +$$

$$+ \frac{\dot{L}(\dot{L}-\dot{N})}{4} = 0 \tag{11}$$

$$R_{01} = -R_{10} = -\frac{\dot{L}}{r} = 0 \tag{12}$$

$$R_{11} = \frac{N''}{2} - \frac{L'N'}{4} + \frac{N'^2}{4} - \frac{L'}{r} +$$

$$+ e^{L-N}\left[\frac{\ddot{L}}{2} + \frac{\dot{L}(\dot{L}-\dot{N})}{4}\right] = 0 \tag{13}$$

$$R_{22} = e^{-L}\left[1 + \frac{1}{2}r(N' - L')\right] - 1 = 0 \tag{14}$$

$$R_{33} = \sin^2\theta\, e^{-L}\left[1 + \frac{1}{2}r(N' - L')\right] +$$

$$- \sin^2\theta = 0 \tag{15}$$

Although these equations look complicated, they can be integrated quite easily. The equation (12) says that L is time independent, and this implies that *all* the terms involving time derivatives in the equations (11) and (13) drop out. The sum of the equation (13) and e^{L-N} times the equation (11) then gives:

$$-\frac{L'}{r} - \frac{N'}{r} = 0 \tag{16}$$

from which we have:

$$N = -L + h(t) \tag{17}$$

where $h(t)$ is an arbitrary function of time [in regard to the space derivatives in the equation (16), this function $h(t)$ behaves like a constant of integration].

When we substitute the equation (17) into the equation (14), we find:

$$e^{-L}\,(-r\,L' + 1) = 1 \tag{18}$$

Since[14] L is independent from time, the partial derivative of L with respect to r is equal to a simple derivative of L with respect to r. Therefore the equation (18) is equal to:

$$e^{-L}\left(-r\,\frac{dL}{dr} + 1\right) = 1 \tag{19}$$

From which we have:

$$-r\,e^{-L}\,\frac{dL}{dr} + e^{-L} = 1 \tag{20}$$

from which we have:

$$-r\,e^{-L}\,\frac{dL}{dr} = 1 - e^{-L} \tag{21}$$

from which we have:

---

[14] From here on, I will modify in a substantially way the treatment of H. C. Ohanian and R. Ruffini: cf. OHANIAN – RUFFINI pp. 293-298.

$$- e^{-L} \frac{dL}{1 - e^{-L}} = \frac{dr}{r} \qquad (22)$$

from which we have:

$$- e^{-L} \frac{dL}{e^{-L}(e^{L} - 1)} = \frac{dr}{r} \qquad (23)$$

from which, finally, we have:

$$\frac{dL}{1 - e^{L}} = \frac{dr}{r} \qquad (24)$$

By integrating the two parts of the equation (24), we have:

$$\int \frac{dL}{1 - e^{L}} = \int \frac{dr}{r} \qquad (25)$$

From which we have:

$$L - \ln(|1 - e^{L}|) = \ln r + K \qquad (26)$$

where K is a constant and in particular a real number since we are operating in the field of the real numbers and not in the field of the imaginary numbers or (in the field) of the complex numbers. In fact we can note that in the equation (25) e^{L} is an element changed of sign of the metric tensor that is expressed by the equation (7), for which e^{L} must be always a real number. On the other hand, r must be always a real number because represents the radius of polar coordinates. Analogously dr must be always a real number because it (dr) is the infinitesimal variation of a real number. Moreover, from the equation (24), because of the fact that e^L, r and dr are all real, we can say that also dL must be real. For which when we integrate the equation (25) the integration is done in the real field both in dL and in dr: the integration constant must, therefore, be a real number. As we will see, the fact that K is a real number is very important.

Therefore, from the equation (26) we have:

$$L - K = \ln r + \ln(|1 - e^L|) \qquad (27)$$

From which we have two possibilities: $1 - e^L > 0$ and $1 - e^L < 0$.

In the case of $1 - e^L > 0$ we have:

$$L - K = \ln\{r\,(1 - e^L)\} \qquad (28)$$

from which we have:

$$e^{-K}\,e^L = r\,(1 - e^L) \qquad (29)$$

from which we have:

$$e^L(e^{-K} + r) = r \qquad (30)$$

from which we have:

$$e^L = \frac{r}{e^{-K} + r} \qquad (31)$$

from which, finally, we have:

$$e^L = \frac{1}{1 + \dfrac{e^{-K}}{r}} \tag{32}$$

where e^{-K} is a positive real number, since e^x is positive for any real number x, and $-K$ is a real number. Moreover, we can see that according to the equation (32) e^L is always < 1 for any value of $r > 0$, for which $1 - e^L$ is always > 0 for any value of $r > 0$. [In fact, we must remember that we have obtained this formula (32) in the hypothesis of $1 - e^L > 0$].

Instead in the case of $1 - e^L < 0$ we have:

$$L - K = \ln\{r\,(e^L - 1)\} \tag{33}$$

from which we have:

$$e^{-K}\, e^L = r\,(e^L - 1) \tag{34}$$

from which we have:

$$e^L(e^{-K} - r) = -r \tag{35}$$

from which we have:

$$e^L = \frac{-r}{e^{-K} - r} \tag{36}$$

from which, finally, we have:

$$e^L = \frac{1}{1 - \dfrac{e^{-K}}{r}} \tag{37}$$

Now this formula (37) is valid, obviously, only in the case of $1 - e^L < 0$ [in fact, we must remember that we have obtained this formula (37) in the hypothesis of $1 - e^L < 0$], that is in the case of $e^L > 1$, and therefore the formula (37) cannot be valid for values of $r < e^{-K}$ because for these values (of r) e^L is < 0 according to the formula (37). On the other hand, we already know that e^{-K} is a positive real number, since e^x is positive for any real number x and $-K$ is a real number.

Since we search a solution that can be valid for all the values of $r > 0$ we must reject the solution (37), because the solution (37) is not valid for all the values of $r > 0$. Therefore, we have that the only acceptable solution is the formula (32). On the other hand, we will see also another reason for discarding the equation (37).

Now we can simplify the equation (32) by means of introducing a positive real constant C, that is equal by definition to e^{-K}:

$$C \equiv e^{-K} \tag{38}$$

Therefore the equation (32) becomes:

$$e^L = \frac{1}{1 + \frac{C}{r}} \tag{39}$$

where C is (a constant that is equal to) a positive real number, in particular C is > 0.

According to the equation (17), the solution for $N(r)$ is then:

$$e^{N(r)} = e^{-L(r)+h(t)} = \left(1 + \frac{C}{r}\right)e^{h(t)} \qquad (40)$$

With the solutions (39) and (40), the space-time interval for the spherically symmetric field takes the form:

$$ds^2 = \left(1 + \frac{C}{r}\right)e^{h(t)}dt^2 - \frac{1}{1+\frac{C}{r}}\,dr^2 - r^2 d\theta^2 +$$

$$- r^2 \sin^2\theta\, d\varphi^2 \qquad (41)$$

where C is (a constant that is equal to) a positive real number, in particular C is > 0.

Here, the time-dependent function $h(t)$ remains undetermined and unknown. The field equations do not determine this function, but we can eliminate it by means of a transformation of the time coordinate. If we adopt a new time coordinate $\tilde{t}$ such that $d\tilde{t} \equiv e^{\frac{h(t)}{2}}dt$, then the function $h(t)$ disappears from view, and the first term of the right side of the equation (41) becomes $\left(1 + \frac{C}{r}\right)d\tilde{t}^2$. We can then omit the tilde, so we obtain:

$$ds^2 = \left(1 + \frac{C}{r}\right)dt^2 - \frac{1}{1+\frac{C}{r}}\,dr^2 - r^2 d\theta^2 +$$

$$- r^2 \sin^2\theta\, d\varphi^2 \qquad (42)$$

Here, we underline again that C is (a constant that is equal to) a positive real number, in particular C is > 0. Commonly instead of C the authors write[15] $- C$ and then consider $- C$ equal to a negative number, in particular to $- 2GM$: now, this is erroneous, as we have seen. In other words, the formula commonly used instead of the expression (42) is not a true solution of the Einstein's field equation of the General Theory of Relativity, if that formula is interpreted in the usual way.

To find the value of C in the equation (42), we must compare this expression (42) with the result obtained from the linear theory. Now, the formula (obtained from the linear theory) commonly used for the comparison is[16]:

$$ds^2 = \left(1 - \frac{2GM}{r}\right) dt^2 - \left(1 + \frac{2GM}{r}\right) (dx^2 +$$

$$+ dy^2 + dz^2) \tag{43}$$

But this formula (obtained from the linear theory) is not expressed as a function of the coordinates measured in the presence of the gravitational fields (that is, measured relatively to a reference frame that is integral with a space-time curved for the presence of the gravitational fields) but as a function of the coordinates measured in the absence of any gravitational field (that is, measured relatively to a reference frame that is integral with a flat space-time)[17], while the expression (42) is expressed as a function of the coordinates measured in the presence of the gravitational fields (that is, measured relatively to a reference frame that is integral with a space-time curved for the presence of the gravitational fields), since the Einstein's field equation of the General Theory of Relativity is expressed as a function of the coordinates measured in the presence of the gravitational fields (that is, measured relatively to a reference frame that is integral

---

[15] Cf. OHANIAN – RUFFINI pp. 293-298.
[16] Cf. OHANIAN – RUFFINI pp. 293-298.
[17] Cf. OHANIAN – RUFFINI pp. 130-131.

with a space-time curved for the presence of the gravitational fields) and we have obtained the expression (42) by starting from the Einstein's field equation of the General Theory of Relativity. Therefore we, instead of the expression (43), need to have the expression (obtained from the linear theory) of the space-time interval, which (space-time interval) be expressed as a function of the coordinates measured in the presence of the gravitational fields (that is, measured relatively to a reference frame that is integral with a space-time curved for the presence of the gravitational fields). In other words, the expression (43) represents the metric that is formally flat when is expressed as a function of the coordinates measured in the presence of the gravitational fields (that is, measured relatively to a reference frame that is integral with a space-time curved for the presence of the gravitational fields) which metric is expressed as a function of the coordinates measured in the absence of any gravitational field (that is, measured relatively to a reference frame that is integral with a flat space-time). Instead we need to have the formula (obtained from the linear theory) which (formula) represent the metric that is formally flat when is expressed as a function of the coordinates measured in the absence of any gravitational field (that is, measured relatively to a reference frame that is integral with a flat space-time), which metric be expressed as a function of the coordinates measured in the presence of the gravitational fields (that is, measured relatively to a reference frame that is integral with a space-time curved for the presence of the gravitational fields).

On the other hand, from the formula (43) we have:

$$d\tau^2 = \left(1 - \frac{2GM}{r}\right) dt^2 \tag{44}$$

where $d\tau$ is the time measured by clocks positioned in the gravitational field (that is, measured relatively to a reference frame that is integral with a space-time curved for the presence of the gravitational field). And therefore for $2GM \ll r$ we have:

$$dt^2 \cong \left(1 + \frac{2GM}{r}\right) d\tau^2 \tag{45}$$

And analogously, from the formula (43) we have relatively to the spatial coordinates:

$$\left(1 + \frac{2GM}{r}\right)(dx^2 + dy^2 + dz^2) =$$

$$= \{dx_g^2 + dy_g^2 + dz_g^2\} =$$

$$= \{dr^2 + r^2 d\theta^2 + r^2 \sin^2 \theta \, d\varphi^2\} \tag{46}$$

And therefore for $2GM \ll r$ we have:

$$(dx^2 + dy^2 + dz^2) =$$

$$\cong \left(1 - \frac{2GM}{r}\right)\{dx_g^2 + dy_g^2 + dz_g^2\} =$$

$$= \left(1 - \frac{2GM}{r}\right)\{dr^2 + r^2 d\theta^2 + r^2 \sin^2 \theta \, d\varphi^2\} \tag{47}$$

where $x_g, \, y_g, \, z_g$ or r, ϑ, φ are the spatial coordinates measured by instrumentation (that is) positioned in the gravitational field (that is, measured relatively to a reference frame that is integral with a space-time curved for the presence of the gravitational field).

Therefore the expression (42) in the limit for $r \to \infty$ must be compared not with the expression (43) but with this expression:

$$ds^2 = \left(1 + \frac{2GM}{r}\right) dt^2 - (1 - \frac{2GM}{r}) (dx^2 +$$

$$+ \, dy^2 + dz^2) \tag{48}$$

or better with this other expression:

$$ds^2 = \left(1 + \frac{2GM}{r}\right) dt^2 - (1 - \frac{2GM}{r}) (dr^2 +$$

$$+ \, r^2 d\theta^2 + r^2 \sin^2 \theta \, d\varphi^2) \tag{49}$$

In fact the formulas (48) and (49), in the case of $\frac{2GM}{r} \ll 1$, represent the expression (obtained from the linear theory) of the space-time interval, which (space-time interval) is expressed as a function of the coordinates measured in the presence of the gravitational fields (that is, measured relatively to a reference frame that is integral with a space-time curved for the presence of the gravitational fields), or in other words represent the expression (obtained from the linear theory) of the metric that is formally flat when is expressed as a function of the coordinates measured in the absence of any gravitational field (that is, measured relatively to a reference frame that is integral with a flat space-time) which metric is expressed as a function of the coordinates measured in the presence of the gravitational fields (that is, measured relatively to a reference frame that is integral with a space-time curved for the presence of the gravitational fields).

Since[18] we cannot compare the expression (42) with the expression (48) or with the expression (49) directly, because they [the expressions (48) and (49)] have a form which is different from that of the expression (42), let us change the equation (42) by introducing a new coordinate $\tilde{r}$:

$$\tilde{r} \equiv \frac{1}{2}\sqrt{r^2 + Cr} + \frac{1}{2}r + \frac{1}{4}C \tag{50}$$

or, equivalently,

$$r \equiv \tilde{r}\left(1 - \frac{C}{4\tilde{r}}\right)^2 \tag{51}$$

This transformation gives:

$$ds^2 = \left(\frac{1 + \frac{C}{4\tilde{r}}}{1 - \frac{C}{4\tilde{r}}}\right)^2 dt^2 - \left(1 - \frac{C}{4\tilde{r}}\right)^4 \{d\tilde{r}^2 +$$

$$+ \tilde{r}^2 d\theta^2 + \tilde{r}^2 \sin^2\theta\, d\varphi^2\} \tag{52}$$

The coordinates used in the expression (52) are called *isotropic*.

In the weak field limit ($\tilde{r} \to \infty$), the equation (52) reduces to:

$$ds^2 = \left(1 + \frac{C}{\tilde{r}}\right)dt^2 - \left(1 - \frac{C}{\tilde{r}}\right)\{d\tilde{r}^2 +$$

---

[18] At this point we return to the usual treatment: cf. OHANIAN – RUFFINI pp. 293-298.

$$+ \; \tilde{r}^2 d\theta^2 + \tilde{r}^2 \sin^2 \theta \, d\varphi^2 \} \tag{53}$$

Obviously, this equation has the same form of the equation (49) [or of the equation (48)]. By comparing the equation (53) with the equation (49), we see that:

$$C = 2GM \tag{54}$$

We can note that this effectively is in accord with the limitation that C must be a positive real number. (The case with $M \to 0$ corresponds to $K \to +\infty$ and $C \equiv e^{-K} \to 0$). On the other hand, this obviously is another reason for discarding the solution (37).

By means of the substitution of C with $2GM$, according to (54), the equation (42) becomes:

$$ds^2 = \left(1 + \frac{2GM}{r}\right) dt^2 - \frac{1}{1 + \frac{2GM}{r}} dr^2 - r^2 d\theta^2 +$$

$$- \; r^2 \sin^2 \theta \, d\varphi^2 \tag{55}$$

where M is (a constant, that is equal to) the total mass of the system.

This is the correct final form of the Schwarzschild solution, that is the correct solution for the case which Schwarzschild has analysed. Instead, the common erroneous expression for the Schwarzschild solution is[19]:

---

[19] Cf. OHANIAN – RUFFINI p. 298.

$$ds^2 = \left(1 - \frac{2GM}{r}\right) dt^2 - \frac{1}{1 - \frac{2GM}{r}} \, dr^2 - r^2 d\theta^2 +$$

$$- r^2 \sin^2 \theta \, d\varphi^2 \tag{56}$$

The difference between the correct formula (55) and the incorrect formula (56) substantially consists in two plus signs instead of two minus signs. As we will see, this entails enormous physical consequences.

2 The behaviour of the clocks positioned in gravitational fields according to the correct Schwarzschild solution

On the other hand, the fact that the coefficient of dt^2 in the equation (55) is > 1 in the presence of a gravitational field entails that in the presence of a gravitational field the clocks go more slowly (that is, that the measurements of time relatively to a reference frame that is integral with a space-time curved for the presence of the gravitational field flow more slowly): and we know that this is in accord with the experimental results. In fact, we have:

$$\left(1 + \frac{2GM}{r}\right) d\tau^2 = dt^2 \tag{57}$$

For which the time τ measured by a clock positioned in the gravitational field (that is, measured relatively to a reference frame that is integral with a space-time curved for the presence of the gravitational field) is less than the time t measured by a clock positioned where there is not any gravitational field (that is, measured relatively to a reference frame that is integral with a flat space-time)[20].

Analogously, the fact that the coefficient of dt^2 in the equation (55) has the form $\left(1 + \frac{2GM}{r}\right)$ entails that the clocks positioned in a more intense gravitational field go more slowly than the clocks positioned in a less intense gravitational field (that is, that the measurements of time relatively to a reference frame that is integral with a space-time curved for the presence of a more intense gravitational field flow more slowly than the measurements of time relatively to a reference frame that is integral with a space-time curved for the presence of a less intense gravitational field). In fact we have:

$$\left(1 + \frac{2GM}{r_1}\right) d\tau_1{}^2 = \left(1 + \frac{2GM}{r_2}\right) d\tau_2{}^2 \tag{58}$$

Now, when the gravitational field is more intense (that is, when r is more small) the coefficient of $d\tau^2$ is greater, for which

---

[20] Of course, it is the clock positioned in a gravitational field to go more slowly than another clock positioned in a place without gravitational fields, not time itself: more precisely, we see the clocks to go more slowly in the presence of gravitational fields, because in these cases we use a reference frame that is integral with a space-time curved for the presence of the gravitational fields. On the other hand, it is physically wrong to use the measurements of a clock for measuring time without considering if the clock is in a gravitational field or not, and the intensity of the gravitational field if it is present, that is without considering if we are using a reference frame that is integral with a flat space-time or a reference frame that is integral with a space-time curved for the presence of the gravitational field, and the value of this curvature of the space-time (that is integral with the used reference frame) if it (this curvature) is present. Indeed, for effecting physical measurements of time we must use only well calibrated clocks and the calibration of the clocks obviously depends on the presence or the absence of a gravitational field, and on the intensity of the gravitational field if it is present, that is depends on the presence or the absence of a curvature in the space-time that is integral with the used reference frame, and on the value of this curvature if it is present (cf. PACE pp. 28-29).

the corresponding time flows more slowly (that is, the measurements of time relatively to a reference frame that is integral with a space-time curved for the presence of a more intense gravitational field flow more slowly than the measurements of time relatively to a reference frame that is integral with a space-time curved for the presence of a less intense gravitational field)[21]. In particular, we have:

$$\frac{d\tau_2}{d\tau_1} = \sqrt{\frac{1 + \frac{2GM}{r_1}}{1 + \frac{2GM}{r_2}}} \tag{59}$$

For which when $r_2 > r_1$, that is when the second gravitational field is less intense that the first gravitational field, then $d\tau_2 > d\tau_1$: we have the well-known result (that has been confirmed by numerous experiments) that the more intense is the gravitational field the more slowly the clocks go (that is, that the measurements of time relatively to a reference frame that is

---

[21] Also here, of course, it is the clock positioned in a gravitational field to go more slowly than another clock positioned in a place without gravitational fields, not time itself: more precisely, we see the clocks to go more slowly in the presence of gravitational fields, because in these cases we use a reference frame that is integral with a space-time curved for the presence of the gravitational fields. On the other hand, it is physically wrong to use the measurements of a clock for measuring time without considering if the clock is in a gravitational field or not, and the intensity of the gravitational field if it is present, that is without considering if we are using a reference frame that is integral with a flat space-time or a reference frame that is integral with a space-time curved for the presence of the gravitational field, and the value of this curvature of the space-time (that is integral with the used reference frame) if it (this curvature) is present. Indeed, for effecting physical measurements of time we must use only well calibrated clocks and the calibration of the clocks obviously depends on the presence or the absence of a gravitational field, and on the intensity of the gravitational field if it is present, that is depends on the presence or the absence of a curvature in the space-time that is integral with the used reference frame, and on the value of this curvature if it is present (cf. PACE pp. 28-29).

integral with a space-time curved for the presence of a more intense gravitational field flow more slowly than the measurements of time relatively to a reference frame that is integral with a space-time curved for the presence of a less intense gravitational field).

Analogously, the ratio of the relative frequencies $v_1 = = 1/d\tau_1$ and $v_2 = 1/d\tau_2$ is:

$$\frac{v_1}{v_2} = \sqrt{\frac{1 + \dfrac{2GM}{r_1}}{1 + \dfrac{2GM}{r_2}}} \tag{60}$$

This is the correct formula for the gravitational redshift of light in the correct Schwarzschild geometry. We can note that the two formulas (59) and (60) are different from those obtained from the incorrect expression of the Schwarzschild solution by means of an incorrect procedure[22], but are equal at the first order in $\dfrac{2GM}{r}$ (to the incorrect formulas). In fact, the formulas commonly used, instead of the two correct formulas (59) and (60), are respectively:

$$\frac{d\tau_2}{d\tau_1} = \sqrt{\frac{1 - \dfrac{2GM}{r_2}}{1 - \dfrac{2GM}{r_1}}} \tag{61}$$

$$\frac{v_1}{v_2} = \sqrt{\frac{1 - \dfrac{2GM}{r_2}}{1 - \dfrac{2GM}{r_1}}} \tag{62}$$

---

[22] Cf. OHANIAN – RUFFINI pp. 293-298.308-309.

It is true that normally $\dfrac{2GM}{r}$ is very small, but at least theoretically we could measure experimentally the difference between the predictions of the correct formulas [(59) and (60)] and the predictions of the incorrect formulas [(61) and (62)].

On the other hand, the commonly used formula (56) can be in accord (at the first order in $\dfrac{2GM}{r}$) with the experimental results of the slowing down of the clocks in the presence of gravitational fields (that is, of the fact that the measurements of time relatively to a reference frame that is integral with a space-time curved for the presence of a more intense gravitational field flow more slowly than the measurements of time relatively to a reference frame that is integral with a space-time curved for the presence of a less intense gravitational field), only if we consider the time t {that is present in it [the formula (56)]} not as the time measured in the gravitational field (that is, measured relatively to a reference frame that is integral with a space-time curved for the presence of the gravitational field), but as the time measured in the absence of gravitational fields (that is, measured relatively to a reference frame that is integral with a flat space-time), as we have seen for the formula of the linear approximation. However, this consideration would be against the Einstein's equations (or more precisely against the Einstein's field equation) of the General Theory of Relativity: in fact we have obtained the correct formula by starting from the Einstein's equations (or more precisely from the Einstein's field equation) of the General Theory of Relativity.

3 The time delay and the time beforehand of light

Normally the time delay of light[23] has been measured by means of coordinates that are relative to instruments (that are) positioned externally to the gravitational field (that is, measured relatively to a reference frame that is integral with a flat space-time) and in the case of $\dfrac{2GM}{r} \ll 1$. Now, in these conditions the formulas (43) and (56) are good approximations, at the first order in $\dfrac{2GM}{r}$, of the correct formulas. Therefore my correction to the Schwarzschild solution does not change the predictions of the theory relatively to these measurements, at least at the first order in $\dfrac{2GM}{r}$, and these phenomena have been measured only at the first order in $\dfrac{2GM}{r}$ [24].

Moreover, the time delay of light has been calculated also as mere application of the linear approximation of the gravitational field[25], that is only by means of the formula (43), and therefore, also for this, my correction to the Schwarzschild solution does not change the predictions of the theory, at least at the first order in $\dfrac{2GM}{r}$. For example, by imposing the usual condition that the propagation of light corresponds to a lightlike interval, that is by imposing $ds^2 = 0$, in the formula (43) we have:

$$0 = \left(1 - \frac{2GM}{r}\right) dt^2 - \left(1 + \frac{2GM}{r}\right)(dx^2 +$$

---

[23] Cf. OHANIAN – RUFFINI pp. 142-149; MISNER – THORNE – WHEELER pp. 1103.1106-1109.

[24] Cf. OHANIAN – RUFFINI pp. 142-149.

[25] Cf. OHANIAN – RUFFINI pp. 142-149.

$$+ \, dy^2 + dz^2) \tag{63}$$

And, therefore, we have:

$$\frac{dx^2 + dy^2 + dz^2}{dt^2} = \frac{1 - \dfrac{2GM}{r}}{1 + \dfrac{2GM}{r}} \tag{64}$$

And then the velocity of light v is:

$$v = \frac{\sqrt{dx^2 + dy^2 + dz^2}}{dt} = \sqrt{\frac{1 - \dfrac{2GM}{r}}{1 + \dfrac{2GM}{r}}} =$$

$$\cong 1 - \frac{2GM}{r} \tag{65}$$

The fact that in the formula (65) v is < 1 (in cgs units, v is $< c$) produces the time delay of light.

Consequently, in this case the accord until now between the experimental results and the theory remains unchanged with my correction of the Schwarzschild solution[26].

On the other hand, if we use the coordinates measured in the gravitational field (that is, measured relatively to a reference frame that is integral with a space-time curved for the presence of the gravitational field), always in the case of $\dfrac{2GM}{r} \ll 1$, then we must use the formula (48) instead of the formula (43) and by imposing the usual condition that the propagation of light

---

[26] Cf. OHANIAN – RUFFINI pp. 142-149.

corresponds to a lightlike interval, that is by imposing $ds^2 = 0$, we have:

$$0 = \left(1 + \frac{2GM}{r}\right) dt^2 - \left(1 - \frac{2GM}{r}\right)(dx^2 + \\ + dy^2 + dz^2) \tag{66}$$

And therefore, we have:

$$\frac{dx^2 + dy^2 + dz^2}{dt^2} = \frac{1 + \frac{2GM}{r}}{1 - \frac{2GM}{r}} \tag{67}$$

And then the velocity of light v is:

$$v = \frac{\sqrt{dx^2 + dy^2 + dz^2}}{dt} = \sqrt{\frac{1 + \frac{2GM}{r}}{1 - \frac{2GM}{r}}} = \\ \cong 1 + \frac{2GM}{r} \tag{68}$$

The fact that in the formula (68) v is > 1 (in cgs units, v is $> c$) produces a time beforehand of light. On the other hand, the fact that the velocity of light is > 1 does not violate the Special Theory of Relativity[27] because in this Theory the velocity of light is always 1 only in vacuum and relatively to inertial reference frames, and in this case we are not measuring in vacuum and

---

[27] Cf. EINSTEIN (1905).

relatively to an inertial reference frame, for the presence of the gravitational field (that is, in this case there is not a vacuum for the presence of the gravitational field and, moreover, we are not measuring relatively to a reference frame that is integral with a flat space-time but we are measuring relatively to a reference frame that is integral with a space-time curved for the presence of the gravitational field). In fact, for example, if we use the non-inertial reference frame that is integral with the surface of the Earth we see that the stars do a whole turn in the sky in one day relatively to this reference frame, and therefore the stars relatively to this non-inertial reference frame move with a velocity much greater than that of light in vacuum relatively to inertial reference frames; but this does not violate the Special Theory of Relativity, since this reference frame (that is integral with the surface of the Earth) is not inertial. This example is valid for all the situations in which we will find, in this book, a velocity of light greater than 1 (in cgs units, greater than c) relatively to a non-inertial reference frame.

Moreover, we can obtain the radial velocity of light in the space-time corresponding to the correct Schwarzschild solution (55) by means of imposing the usual condition that the propagation of light corresponds to a lightlike interval, that is by means of imposing $ds^2 = 0$, and by means of imposing the conditions for a radial motion: $d\theta = 0$ and $d\varphi = 0$:

$$0 = \left(1 + \frac{2GM}{r}\right) dt^2 - \frac{1}{1 + \frac{2GM}{r}} \, dr^2 \tag{69}$$

Therefore, we have:

$$\frac{dr^2}{dt^2} = \left(1 + \frac{2GM}{r}\right)^2 \tag{70}$$

And then the radial velocity of light v is:

$$v = \frac{dr}{dt} = 1 + \frac{2GM}{r} \tag{71}$$

Also here the fact that in the formula (71) v is > 1 (in cgs units, v is $> c$) produces a time beforehand of light. On the other hand, also here the fact that the velocity of light is > 1 does not violate the Special Theory of Relativity[28] because in this Theory the velocity of light is always 1 only in vacuum and relatively to inertial reference frames, and in this case we are not measuring in vacuum and relatively to an inertial reference frame, for the presence of the gravitational field (that is, in this case there is not a vacuum for the presence of the gravitational field and, moreover, we are not measuring relatively to a reference frame that is integral with a flat space-time but we are measuring relatively to a reference frame that is integral with a space-time curved for the presence of the gravitational field).

On the other hand, if we proceed in an analogous way to the linear case also in the non-linear case, by means of a proceeding that is the inverse of that with which we have transformed the expression (43) in the expression (48) or in the expression (49), for having the space-time interval expressed as a function of the coordinates t_i and r_i of an inertial reference frame, we obtain that in the expression of the space-time interval the coefficients of dt_i^2 and dr_i^2 are the inverses respectively of those of dt^2 and dr^2 in the expression (55), for which the radial velocity of light v_i relatively to an inertial reference frame results equal to:

$$v_i = \frac{dr_i}{dt_i} = \frac{1}{1 + \frac{2GM}{r}} \tag{72}$$

---

[28] Cf. EINSTEIN (1905).

that is to the inverse of the value in the non-inertial reference frame. Therefore, correctly, the value of v_i is always ≤ 1. Of course, in the formula (72) r is still relative to the non-inertial reference frame, but we can think of replacing [in the formula (72)] r point by point with the corresponding value of the inertial reference frame. However we have that, since $v_i = 1/v$, v_i can be ≤ 1, as the Special Theory of Relativity requires[29], only if $v \geq 1$. Therefore the formula (72) is in accord with the Special Theory of Relativity exactly because the formula (71) affirms that $v \geq 1$.

4 The perihelion precession and the deflection of light

As for the perihelion precession[30] and the deflection of light[31], we have that these phenomena have been measured by means of coordinates that are relative to instruments positioned externally to the gravitational field (that is, measured relatively to a reference frame that is integral with a flat space-time) and in the case of $\dfrac{2GM}{r} \ll 1$. Now, in these conditions the formulas (43) and (56) are good approximations, at the first order in $\dfrac{2GM}{r}$, of the correct formulas. Therefore my correction to the Schwarzschild solution does not change the predictions of the theory relatively to these measurements, at least at the first order

---

[29] Cf. EINSTEIN (1905).
[30] Cf. EINSTEIN (1916); OHANIAN – RUFFINI pp. 299-305; MISNER – THORNE – WHEELER pp. 391.433.1110-1116; WEINBERG (1972) pp. 194-201.230-233; ANDERSON pp. 404-410.
[31] Cf. EINSTEIN (1916); OHANIAN – RUFFINI pp. 138-142; MISNER – THORNE – WHEELER pp. 184-185.446.679.1101-1105; WEINBERG (1972) pp. 188-194; ANDERSON pp. 410-414.

in $\dfrac{2GM}{r}$, and these phenomena have been measured only at the first order in $\dfrac{2GM}{r}$ [32].

Moreover, the deflection of light has been calculated also as mere application of the linear approximation of the gravitational field[33], that is only by means of the formula (43), and therefore, also for this, my correction to the Schwarzschild solution does not change the predictions of the theory, at least at the first order in $\dfrac{2GM}{r}$.

Consequently, also in these cases (the perihelion precession and the deflection of light) the accord until now between the experimental results and the theory remains unchanged with my correction of the Schwarzschild solution[34].

5 There is not any event horizon, and therefore there is not any black hole, in the correct Schwarzschild solution

We have seen that the correct Schwarzschild solution is:

$$ds^2 = \left(1 + \frac{2GM}{r}\right) dt^2 - \frac{1}{1 + \frac{2GM}{r}} dr^2 - r^2 d\theta^2 +$$

$$- r^2 \sin^2 \theta \, d\varphi^2 \tag{73}$$

---

[32] Cf. OHANIAN – RUFFINI pp. 138-142.299-305.

[33] Cf. OHANIAN – RUFFINI pp. 138-142.

[34] Cf. OHANIAN – RUFFINI pp. 138-142.299-305.

Therefore, contrary to the erroneous solution (56), there is not any singularity at the Schwarzschild radius $r_s \equiv 2GM$ (in cgs units, $r_s \equiv \dfrac{2GM}{c^2}$). In fact at this radius we have:

$$ds^2 \; = \; 2dt^2 \, - \, \frac{1}{2} \, dr^2 \, - \, r^2 d\theta^2 \, - \, r^2 \sin^2 \theta \, d\varphi^2 \qquad (74)$$

where we do not see any singularity. In particular, for the formula (59) a clock positioned at $r = r_s$ goes more slowly than another equal clock positioned at $r = +\infty$ by a factor $\sqrt{2}$ [35]. On the other hand, the r_s is a characteristic value, since it is the value for which the coefficient of dt^2 is equal to 2 (which is twice the value that this coefficient has in the absence of gravitational fields) and the coefficient of dr^2 is equal to $-\dfrac{1}{2}$ (which is the half of the value that this coefficient has in the absence of gravitational fields).

----

[35] Also here, of course, it is the clock positioned in a gravitational field to go more slowly than another clock positioned in a place without gravitational fields, not time itself: more precisely, we see the clocks to go more slowly in the presence of gravitational fields, because in these cases we use a reference frame that is integral with a space-time curved for the presence of the gravitational fields. On the other hand, it is physically wrong to use the measurements of a clock for measuring time without considering if the clock is in a gravitational field or not, and the intensity of the gravitational field if it is present, that is without considering if we are using a reference frame that is integral with a flat space-time or a reference frame that is integral with a space-time curved for the presence of the gravitational field, and the value of this curvature of the space-time (that is integral with the used reference frame) if it (this curvature) is present. Indeed, for effecting physical measurements of time we must use only well calibrated clocks and the calibration of the clocks obviously depends on the presence or the absence of a gravitational field, and on the intensity of the gravitational field if it is present, that is depends on the presence or the absence of a curvature in the space-time that is integral with the used reference frame, and on the value of this curvature if it is present (cf. PACE pp. 28-29).

However, we have not any singularity in the expression (73) for any value of $r > 0$. Therefore we can say that there is not any event horizon.

In this expression (73) only at $r = 0$ we have a singularity in the case of $M > 0$: in fact in this case for $r \to 0$ the coefficient of dt^2 tends to $+\infty$ and the coefficient of dr^2 tends to zero. But this space-time interval has been calculated for the space surrounding a central mass; now, if $r = 0$ were considered external to the central mass, then the central mass would not have space for existing, for which we must consider only values of $r > 0$ if $M > 0$. Moreover, in the case in which the central mass M is equal to zero we have not any singularity in the expression (73) at $r = 0$. Therefore the expression (73) has not any singularity in its field of application.

Furthermore, the redshift is always not infinite for any value of $r > 0$, in particular this is true for $r = r_s$, contrary to the common treatment of the Schwarzschild solution[36].

On the other hand, the coefficient of dt^2 is always ≥ 1, therefore since the coefficient of dt^2 is always not negative the time (that is, the temporal coordinate) does not become in any case as a spatial coordinate, contrary to the common treatment of the space-time inside the event horizon[37]. Moreover, since the coefficient of dr^2 is always not positive the spatial coordinate r does not become in any case as a temporal coordinate, contrary to the common treatment of the space-time inside the event horizon[38].

Therefore the light cones are always orientated in the usual way, in particular there is not any horizontal inclination of the

---

[36] Cf. OHANIAN – RUFFINI pp. 293-298.324-342; MISNER – THORNE – WHEELER pp. 819-940.

[37] Cf. OHANIAN – RUFFINI pp. 324-342; MISNER – THORNE – WHEELER pp. 819-940.

[38] Cf. OHANIAN – RUFFINI pp. 324-342; MISNER – THORNE – WHEELER pp. 819-940.

light cones, contrary to the common treatment of the space-time inside the event horizon[39].

Furthermore, since every black hole implies the existence of an event horizon[40], we can say that in the correct Schwarzschild solution there is not any possibility of existence of a black hole, because, as we have seen, there is not any possibility of existence of an event horizon in the correct Schwarzschild solution.

---

[39] Cf. OHANIAN – RUFFINI pp. 324-342; MISNER – THORNE – WHEELER pp. 819-940.

[40] Cf. OHANIAN – RUFFINI pp. 324-342; MISNER – THORNE – WHEELER pp. 819-940; WEINBERG (1972) pp. 207-208.297.349.479.

CHAPTER II
OTHER CORRECT SOLUTIONS
OF THE EINSTEIN'S FIELD
EQUATION:
THAT OF KERR,
THAT OF REISSNER-NORDSTRØM
AND THAT OF KERR-NEWMAN

1 The correct Kerr solution

By analogy with the case of the Schwarzschild solution we can extend what we have seen for the case of this solution to other similar solutions of the nonlinear Einstein's equations (or more precisely of the Einstein's field equation) of the General Theory of Relativity, in particular to the Kerr solution[41], to the Reissner-Nordstrøm solution[42] and to the Kerr-Newman solution[43], which solutions are extensions of the Schwarzschild solution.

As for the Kerr solution, that is the solution that represents the curved space-time geometry surrounding a rotating mass, we have that the commonly used expression is[44]:

---

[41] Cf. OHANIAN – RUFFINI pp. 343-346.349-360; MISNER – THORNE – WHEELER pp. 877-878; WEINBERG (1972) pp. 240.349.

[42] Cf. OHANIAN – RUFFINI pp. 346-348.360; MISNER – THORNE – WHEELER pp. 840-841.920-921.

[43] Cf. MISNER – THORNE – WHEELER pp. 875-885.891-894; OHANIAN – RUFFINI pp. 348. 355.

[44] Cf. OHANIAN – RUFFINI p. 343.

$$ds^2 = dt^2 - \frac{\rho^2}{\Delta} dr^2 - \rho^2 d\theta^2 +$$

$$- (r^2 + a^2)\, \sin^2 \theta\, d\varphi^2 +$$

$$- \frac{2GMr}{\rho^2} \{dt - a \sin^2 \theta\, d\varphi\}^2 \qquad (75)$$

where ρ^2 and Δ are functions of r and θ:

$$\rho^2 \equiv r^2 + a^2 \cos^2 \theta \qquad (76)$$

$$\Delta \equiv r^2 - 2GMr + a^2 \qquad (77)$$

and M and a are constants: M is the total mass of the system and a is the spin angular momentum of the system per unit mass.

But this expression is erroneous: in fact, by means of the fact that the correct Kerr solution must be equal to the correct Schwarzschild solution for $a = 0$, we can say that the correct Kerr solution is expressed by this other expression:

$$ds^2 = dt^2 - \frac{\rho^2}{\Delta} dr^2 - \rho^2 d\theta^2 +$$

$$- (r^2 + a^2)\, \sin^2 \theta\, d\varphi^2 +$$

$$+ \frac{2GMr}{\rho^2} \{dt - a \sin^2 \theta\, d\varphi\}^2 \qquad (78)$$

where ρ^2 and Δ are functions of r and θ:

$$\rho^2 \equiv r^2 + a^2 \cos^2 \theta \tag{79}$$

$$\Delta \equiv r^2 + 2GMr + a^2 \tag{80}$$

and M and a are constants: M is the total mass of the system and a is the spin angular momentum of the system per unit mass.

Also in this case as in that of Schwarzschild, we have not any singularity in the correct expression (78) for any value of $r > 0$. Therefore we can say that there is not any event horizon and therefore that there is not any black hole[45].

Furthermore, the redshift is always not infinite for any value of $r > 0$, as in the correct Schwarzschild solution.

In particular, also here the coefficient of dt^2 is always ≥ 1, therefore since the coefficient of dt^2 is always not negative the time (that is, the temporal coordinate) does not become in any case as a spatial coordinate, contrary to the common treatment of the space-time inside the event horizon[46]. Moreover, since the coefficient of dr^2 is always not positive, also here the spatial coordinate r does not become in any case as a temporal coordinate, contrary to the common treatment of the space-time inside the event horizon[47].

Therefore also here the light cones are always orientated in the usual way, in particular there is not any horizontal inclination of the light cones, contrary to the common treatment of the space-time inside the event horizon[48].

---

[45] Cf. OHANIAN – RUFFINI pp. 324-360; MISNER – THORNE – WHEELER pp. 819-940.

[46] Cf. OHANIAN – RUFFINI pp. 324-360; MISNER – THORNE – WHEELER pp. 819-940.

[47] Cf. OHANIAN – RUFFINI pp. 324-360; MISNER – THORNE – WHEELER pp. 819-940.

[48] Cf. OHANIAN – RUFFINI pp. 324-360; MISNER – THORNE – WHEELER pp. 819-940.

In this case the formulas (59), (60) and (71) for $\sin^2 \theta = 0$ (and therefore for $\cos^2 \theta = 1$) become respectively:

$$\frac{d\tau_2}{d\tau_1} = \sqrt{\frac{1 + \dfrac{2GMr_1}{r_1^2 + a^2}}{1 + \dfrac{2GMr_2}{r_2^2 + a^2}}} \qquad (81)$$

$$\frac{v_1}{v_2} = \sqrt{\frac{1 + \dfrac{2GMr_1}{r_1^2 + a^2}}{1 + \dfrac{2GMr_2}{r_2^2 + a^2}}} \qquad (82)$$

$$v = \frac{dr}{dt} = \sqrt{\frac{\Delta}{\rho^2}\left(1 + \frac{2GMr}{\rho^2}\right)} =$$

$$= \sqrt{\frac{r^2 + 2GMr + a^2}{r^2 + a^2}\left(1 + \frac{2GMr}{r^2 + a^2}\right)} =$$

$$= 1 + \frac{2GMr}{r^2 + a^2} \qquad (83)$$

Therefore also in this case we have that in the presence of a gravitational field the clocks go more slowly (that is, that the measurements of time relatively to a reference frame that is integral with a space-time curved for the presence of a gravitational field flow more slowly). And the more intense is the gravitational field the more slowly the clocks go (that is, the measurements of time relatively to a reference frame that is integral with a space-time curved for the presence of a more intense gravitational field flow more slowly than the measurements of time relatively to a reference frame that is integral with a space-time curved for the presence of a less intense gravitational field)[49].

Moreover, also here the fact that in the formula (83) v is $>$ 1 (in cgs units, v is $> c$) produces a time beforehand of light. On the other hand, also here the fact that the velocity of light is > 1 does not violate the Special Theory of Relativity[50] because in this Theory the velocity of light is always 1 only in vacuum and relatively to inertial reference frames, and in this case we are not measuring in vacuum and relatively to an inertial reference frame, for the presence of the gravitational field (that is, in this case there is not a vacuum for the presence of the gravitational field and, moreover, we are not measuring relatively to a reference frame that is integral with a flat space-time but we are measuring relatively to a reference frame that is integral with a space-time curved for the presence of the gravitational field).

---

[49] Also here, of course, it is the clock positioned in a gravitational field to go more slowly than another clock positioned in a place without gravitational fields, not time itself: more precisely, we see the clocks to go more slowly in the presence of gravitational fields, because in these cases we use a reference frame that is integral with a space-time curved for the presence of the gravitational fields. On the other hand, it is physically wrong to use the measurements of a clock for measuring time without considering if the clock is in a gravitational field or not, and the intensity of the gravitational field if it is present, that is without considering if we are using a reference frame that is integral with a flat space-time or a reference frame that is integral with a space-time curved for the presence of the gravitational field, and the value of this curvature of the space-time (that is integral with the used reference frame) if it (this curvature) is present. Indeed, for effecting physical measurements of time we must use only well calibrated clocks and the calibration of the clocks obviously depends on the presence or the absence of a gravitational field, and on the intensity of the gravitational field if it is present, that is depends on the presence or the absence of a curvature in the space-time that is integral with the used reference frame, and on the value of this curvature if it is present (cf. PACE pp. 28-29).
[50] Cf. EINSTEIN (1905).

On the other hand, also here if we proceed by means of a proceeding that is the inverse of that with which we have transformed the expression (43) in the expression (48) or in the expression (49), for having the space-time interval expressed as a function of the coordinates t_i and r_i of an inertial reference frame, we obtain that in the expression of the space-time interval the coefficients of dt_i^2 and dr_i^2 are the inverses respectively of those of dt^2 and dr^2 in the expression (78), for which the radial velocity of light v_i relatively to an inertial reference frame results equal to:

$$v_i = \frac{dr_i}{dt_i} = \frac{1}{1 + \dfrac{2GMr}{r^2 + a^2}} \tag{84}$$

that is to the inverse of the value in the non-inertial reference frame. Therefore, correctly, the value of v_i is always ≤ 1. Of course, in the formula (84) r is still relative to the non-inertial reference frame, but we can think of replacing [in the formula (84)] r point by point with the corresponding value of the inertial reference frame. However we have that, since $v_i = 1/v$, v_i can be ≤ 1, as the Special Theory of Relativity requires[51], only if $v \geq 1$. Therefore the formula (84) is in accord with the Special Theory of Relativity exactly because the formula (83) affirms that $v \geq 1$.

---

[51] Cf. EINSTEIN (1905).

2 The correct Reissner-Nordstrøm solution

As for the Reissner-Nordstrøm solution[52], that is the solution that represents the curved space-time geometry surrounding an electrically charged mass, we have that the commonly used expressions for the space-time interval and the electric field are respectively[53]:

$$ds^2 = \left(1 - \frac{2GM}{r} + \frac{GQ^2}{r^2}\right) dt^2 +$$

$$- \frac{1}{1 - \frac{2GM}{r} + \frac{GQ^2}{r^2}} dr^2 - r^2 d\theta^2 +$$

$$- r^2 \sin^2 \theta \, d\varphi^2 \tag{85}$$

and

$$E(r) = \frac{Q}{r^2} \tag{86}$$

where M and Q are two constants: M is the total mass of the system and Q is the total electric charge of the system.

But the expression (85) is erroneous: in fact, by means of the fact that the correct Reissner-Nordstrøm solution must be equal to the correct Schwarzschild solution for $Q = 0$, we can

---

say that the correct Reissner-Nordstrøm solution is expressed by this other expression:

$$ds^2 = \left(1 + \frac{2GM}{r} + \frac{GQ^2}{r^2}\right) dt^2 +$$

$$- \frac{1}{1 + \dfrac{2GM}{r} + \dfrac{GQ^2}{r^2}} dr^2 - r^2 d\theta^2 +$$

$$- r^2 \sin^2\theta \, d\varphi^2 \tag{87}$$

where M and Q are two constants: M is the total mass of the system and Q is the total electric charge of the system.

Also in this case as in that of Schwarzschild, we have not any singularity in the correct expression (87) for any value of $r >$ 0. Therefore we can say that there is not any event horizon and therefore that there is not any black hole[54].

Furthermore, the redshift is always not infinite for any value of $r > 0$, as in the correct Schwarzschild solution.

In particular, also here the coefficient of dt^2 is always ≥ 1, therefore since the coefficient of dt^2 is always not negative the time (that is, the temporal coordinate) does not become in any case as a spatial coordinate, contrary to the common treatment of the space-time inside the event horizon[55]. Moreover, since the coefficient of dr^2 is always not positive, also here the spatial coordinate r does not become in any case as a temporal

---

[54] Cf. OHANIAN – RUFFINI pp. 324-360; MISNER – THORNE – WHEELER pp. 819-940.
[55] Cf. OHANIAN – RUFFINI pp. 324-360; MISNER – THORNE – WHEELER pp. 819-940.

coordinate, contrary to the common treatment of the space-time inside the event horizon[56].

Therefore, also here the light cones are always orientated in the usual way, in particular there is not any horizontal inclination of the light cones, contrary to the common treatment of the space-time inside the event horizon[57].

In this case the formulas (59), (60) and (71) become respectively:

$$\frac{d\tau_2}{d\tau_1} = \sqrt{\frac{1 + \frac{2GM}{r_1} + \frac{GQ^2}{r_1^2}}{1 + \frac{2GM}{r_2} + \frac{GQ^2}{r_2^2}}} \tag{88}$$

$$\frac{v_1}{v_2} = \sqrt{\frac{1 + \frac{2GM}{r_1} + \frac{GQ^2}{r_1^2}}{1 + \frac{2GM}{r_2} + \frac{GQ^2}{r_2^2}}} \tag{89}$$

$$v = \frac{dr}{dt} = 1 + \frac{2GM}{r} + \frac{GQ^2}{r^2} \tag{90}$$

Therefore also in this case we have that in the presence of a gravitational field the clocks go more slowly (that is, that the measurements of time relatively to a reference frame that is integral with a space-time curved for the presence of a gravitational field flow more slowly). And the more intense is the

---

[56] Cf. OHANIAN – RUFFINI pp. 324-360; MISNER – THORNE – WHEELER pp. 819-940.
[57] Cf. OHANIAN – RUFFINI pp. 324-360; MISNER – THORNE – WHEELER pp. 819-940.

gravitational field the more slowly the clocks go (that is, the measurements of time relatively to a reference frame that is integral with a space-time curved for the presence of a more intense gravitational field flow more slowly than the measurements of time relatively to a reference frame that is integral with a space-time curved for the presence of a less intense gravitational field)[58].

Moreover, also here the fact that in the formula (90) v is > 1 (in cgs units, v is $> c$) produces a time beforehand of light. On the other hand, also here the fact that the velocity of light is > 1 does not violate the Special Theory of Relativity[59] because in this Theory the velocity of light is always 1 only in vacuum and relatively to inertial reference frames, and in this case we are not measuring in vacuum and relatively to an inertial reference frame, for the presence of the gravitational field (and also of the electromagnetic field) [that is, in this case there is not a vacuum for the presence of the gravitational field (and also of the electromagnetic field) and, moreover, we are not measuring

---

[58] Also here, of course, it is the clock positioned in a gravitational field to go more slowly than another clock positioned in a place without gravitational fields, not time itself: more precisely, we see the clocks to go more slowly in the presence of gravitational fields, because in these cases we use a reference frame that is integral with a space-time curved for the presence of the gravitational fields. On the other hand, it is physically wrong to use the measurements of a clock for measuring time without considering if the clock is in a gravitational field or not, and the intensity of the gravitational field if it is present, that is without considering if we are using a reference frame that is integral with a flat space-time or a reference frame that is integral with a space-time curved for the presence of the gravitational field, and the value of this curvature of the space-time (that is integral with the used reference frame) if it (this curvature) is present. Indeed, for effecting physical measurements of time we must use only well calibrated clocks and the calibration of the clocks obviously depends on the presence or the absence of a gravitational field, and on the intensity of the gravitational field if it is present, that is depends on the presence or the absence of a curvature in the space-time that is integral with the used reference frame, and on the value of this curvature if it is present (cf. PACE pp. 28-29).

[59] Cf. EINSTEIN (1905).

relatively to a reference frame that is integral with a flat space-time but we are measuring relatively to a reference frame that is integral with a space-time curved for the presence of the gravitational field (and also of the electromagnetic field)].

On the other hand, also here if we proceed by means of a proceeding that is the inverse of that with which we have transformed the expression (43) in the expression (48) or in the expression (49), for having the space-time interval expressed as a function of the coordinates t_i and r_i of an inertial reference frame, we obtain that in the expression of the space-time interval the coefficients of dt_i^2 and dr_i^2 are the inverses respectively of those of dt^2 and dr^2 in the expression (87), for which the radial velocity of light v_i relatively to an inertial reference frame results equal to:

$$v_i = \frac{dr_i}{dt_i} = \frac{1}{1 + \frac{2GM}{r} + \frac{GQ^2}{r^2}} \tag{91}$$

that is to the inverse of the value in the non-inertial reference frame. Therefore, correctly, the value of v_i is always ≤ 1. Of course, in the formula (91) r is still relative to the non-inertial reference frame, but we can think of replacing [in the formula (91)] r point by point with the corresponding value of the inertial reference frame. However we have that, since $v_i = 1/v$, v_i can be ≤ 1, as the Special Theory of Relativity requires[60], only if $v \geq 1$. Therefore the formula (91) is in accord with the Special Theory of Relativity exactly because the formula (90) affirms that $v \geq 1$.

---

[60] Cf. EINSTEIN (1905).

3 The correct Kerr-Newman solution

As for the Kerr-Newman solution[61], that is the solution that represents the curved space-time geometry surrounding an electrically charged rotating mass, we have that the commonly used expression is[62]:

$$ds^2 = \frac{\Delta}{\rho^2}[dt - a \sin^2 \theta \, d\varphi]^2 +$$

$$- \frac{\sin^2 \theta}{\rho^2}[(r^2 + a^2)d\varphi - adt]^2 +$$

$$- \frac{\rho^2}{\Delta}dr^2 - \rho^2 d\theta^2 \qquad (92)$$

where ρ^2 and Δ are functions of r and θ:

$$\Delta \equiv r^2 - 2GMr + a^2 + GQ^2 \qquad (93)$$

$$\rho^2 \equiv r^2 + a^2 \cos^2 \theta \qquad (94)$$

and M, Q and a are constants: M is the total mass of the system, Q is the total electric charge of the system, and a is the angular momentum of the system per unit mass.

But also here the expression (92) is erroneous: in fact, by means of the fact that the correct Kerr-Newman solution must be equal to the correct Kerr solution for $Q = 0$, and must be equal to the correct Reissner-Nordstrøm solution for $a = 0$, and must be

---

[61] Cf. MISNER – THORNE – WHEELER pp. 875-885.891-894; OHANIAN – RUFFINI pp. 348.355.
[62] Cf. MISNER – THORNE – WHEELER p. 877.

equal to the correct Schwarzschild solution for $Q = 0$ and $a = 0$, we can say that the correct Kerr-Newman solution is expressed by this other expression:

$$ds^2 = \frac{\Delta}{\rho^2}[dt - a \sin^2 \theta \, d\varphi]^2 +$$

$$- \frac{\sin^2 \theta}{\rho^2}[(r^2 + a^2)d\varphi - adt]^2 +$$

$$- \frac{\rho^2}{\Delta}dr^2 - \rho^2 d\theta^2 \tag{95}$$

where ρ^2 and Δ are functions of r and θ:

$$\Delta \equiv r^2 + 2GMr + a^2 + GQ^2 \tag{96}$$

$$\rho^2 \equiv r^2 + a^2 \cos^2 \theta \tag{97}$$

and M, Q and a are constants: M is the total mass of the system, Q is the total electric charge of the system, and a is the angular momentum of the system per unit mass.

As we can see, the difference between the correct formula and the incorrect formula is only in the definition of Δ, in which we have respectively $+ 2GMr$ instead of $- 2GMr$.

Also in this case as in that of Schwarzschild, we have not any singularity in the correct expression (95) for any value of $r > 0$. Therefore we can say that there is not any event horizon and therefore that there is not any black hole[63].

---

[63] Cf. OHANIAN – RUFFINI pp. 324-360; MISNER – THORNE – WHEELER pp. 819-940.

Furthermore, the redshift is always not infinite for any value of $r > 0$, as in the correct Schwarzschild solution.

In particular, also here the coefficient of dt^2 is always ≥ 1, therefore since the coefficient of dt^2 is always not negative the time (that is, the temporal coordinate) does not become in any case as a spatial coordinate, contrary to the common treatment of the space-time inside the event horizon[64]. Moreover, since the coefficient of dr^2 is always not positive, also here the spatial coordinate r does not become in any case as a temporal coordinate, contrary to the common treatment of the space-time inside the event horizon[65].

Therefore, also here the light cones are always orientated in the usual way, in particular there is not any horizontal inclination of the light cones, contrary to the common treatment of the space-time inside the event horizon[66].

In this case the formulas (59), (60) and (71) for $\sin^2 \theta = 0$ (and therefore for $\cos^2 \theta = 1$) become respectively:

$$\frac{d\tau_2}{d\tau_1} = \sqrt{\frac{1 + \dfrac{2GMr_1 + GQ^2}{r_1^2 + a^2}}{1 + \dfrac{2GMr_2 + GQ^2}{r_2^2 + a^2}}} \tag{98}$$

---

[64] Cf. OHANIAN – RUFFINI pp. 324-360; MISNER – THORNE – WHEELER pp. 819-940.

[65] Cf. OHANIAN – RUFFINI pp. 324-360; MISNER – THORNE – WHEELER pp. 819-940.

[66] Cf. OHANIAN – RUFFINI pp. 324-360; MISNER – THORNE – WHEELER pp. 819-940.

$$\frac{v_1}{v_2} = \sqrt{\frac{1 + \dfrac{2GMr_1 + GQ^2}{r_1^2 + a^2}}{1 + \dfrac{2GMr_2 + GQ^2}{r_2^2 + a^2}}} \tag{99}$$

$$v = \frac{dr}{dt} = 1 + \frac{2GMr + GQ^2}{r^2 + a^2} \tag{100}$$

Therefore also in this case we have that in the presence of a gravitational field the clocks go more slowly (that is, that the measurements of time relatively to a reference frame that is integral with a space-time curved for the presence of a gravitational field flow more slowly). And the more intense is the gravitational field the more slowly the clocks go (that is, the measurements of time relatively to a reference frame that is integral with a space-time curved for the presence of a more intense gravitational field flow more slowly than the measurements of time relatively to a reference frame that is integral with a space-time curved for the presence of a less intense gravitational field)[67].

---

[67] Also here, of course, it is the clock positioned in a gravitational field to go more slowly than another clock positioned in a place without gravitational fields, not time itself: more precisely, we see the clocks to go more slowly in the presence of gravitational fields, because in these cases we use a reference frame that is integral with a space-time curved for the presence of the gravitational fields. On the other hand, it is physically wrong to use the measurements of a clock for measuring time without considering if the clock is in a gravitational field or not, and the intensity of the gravitational field if it is present, that is without considering if we are using a reference frame that is integral with a flat space-time or a reference frame that is integral with a space-time curved for the presence of the gravitational field, and the value of this curvature of the space-time (that is integral with the used reference frame) if it (this curvature) is present. Indeed, for effecting physical measurements of time we must use only well calibrated clocks and the calibration of the clocks obviously depends on the presence or the absence of a gravitational field, and on the intensity of the gravitational field if it is present, that is depends on the presence

Moreover, also here the fact that in the formula (100) v is $>$ 1 (in cgs units, v is $> c$) produces a time beforehand of light. On the other hand, also here the fact that the velocity of light is > 1 does not violate the Special Theory of Relativity[68] because in this Theory the velocity of light is always 1 only in vacuum and relatively to inertial reference frames, and in this case we are not measuring in vacuum and relatively to an inertial reference frame, for the presence of the gravitational field (and also of the electromagnetic field) [that is, in this case there is not a vacuum for the presence of the gravitational field (and also of the electromagnetic field) and, moreover, we are not measuring relatively to a reference frame that is integral with a flat space-time but we are measuring relatively to a reference frame that is integral with a space-time curved for the presence of the gravitational field (and also of the electromagnetic field)].

On the other hand, also here if we proceed by means of a proceeding that is the inverse of that with which we have transformed the expression (43) in the expression (48) or in the expression (49), for having the space-time interval expressed as a function of the coordinates t_i and r_i of an inertial reference frame, we obtain that in the expression of the space-time interval the coefficients of dt_i^2 and dr_i^2 are the inverses respectively of those of dt^2 and dr^2 in the expression (95), for which the radial velocity of light v_i relatively to an inertial reference frame results equal to:

$$v_i = \frac{dr_i}{dt_i} = \frac{1}{1 + \dfrac{2GMr + GQ^2}{r^2 + a^2}} \tag{101}$$

---

[68] Cf. EINSTEIN (1905).

that is to the inverse of the value in the non-inertial reference frame. Therefore, correctly, the value of v_i is always ≤ 1. Of course, in the formula (101) r is still relative to the non-inertial reference frame, but we can think of replacing [in the formula (101)] r point by point with the corresponding value of the inertial reference frame. However we have that, since $v_i = 1/v$, v_i can be ≤ 1, as the Special Theory of Relativity requires[69], only if $v \geq 1$. Therefore the formula (101) is in accord with the Special Theory of Relativity exactly because the formula (100) affirms that $v \geq 1$.

---

[69] Cf. EINSTEIN (1905).

CHAPTER III
THE CORRECT SOLUTIONS OF SCHWARZSCHILD, OF KERR, OF REISSNER-NORDSTRØM AND OF KERR-NEWMAN IN THE CASE IN WHICH THE COSMOLOGICAL CONSTANT IS GREATER THAN ZERO

1 The correct Schwarzschild solution in the case in which the cosmological constant is greater than zero

In the case in which the cosmological constant Λ is > 0 (instead of being equal to zero), the Einstein's field equation (of the General Theory of Relativity) in vacuum is[70]:

$$R_{\mu\nu} - \frac{1}{2} g_{\mu\nu} R = - \Lambda g_{\mu\nu} \qquad (102)$$

instead of :

$$R_{\mu\nu} - \frac{1}{2} g_{\mu\nu} R = 0 \qquad (103)$$

---

[70] Cf. OHANIAN – RUFFINI p. 299; WEINBERG (1972) p. 613.

In the case in which the cosmological constant Λ is greater than zero the Schwarzschild solution is commonly written as[71]:

$$ds^2 = \left(1 - \frac{2GM}{r} - \frac{\Lambda r^2}{3}\right) dt^2 +$$

$$- \frac{1}{1 - \frac{2GM}{r} - \frac{\Lambda r^2}{3}} dr^2 +$$

$$- r^2 d\theta^2 - r^2 \sin^2 \theta \, d\varphi^2 \qquad (104)$$

Because Λ is very small there is a range of radii (in cgs units, $\frac{GM}{c^2} \ll r \ll \frac{c}{\sqrt{\Lambda}}$) in which the space-time geometry is nearly flat. At the lower end of this range, the effect of the mass M dominates; at the upper end of this range, the effect of the cosmological term dominates. Moreover, in the Newtonian approximation the Newtonian potential becomes:

$$\Phi = - \frac{GM}{r} - \frac{\Lambda}{6} r^2 \qquad (105)$$

Now, by analogy with the derivation of the correct Schwarzschild solution in the case with $\Lambda = 0$, we have that the correct Schwarzschild solution in the case with $\Lambda > 0$ is:

$$ds^2 = \left(1 + \frac{2GM}{r} + \frac{\Lambda r^2}{3}\right) dt^2 +$$

---

[71] Also here I will correct the treatment that was made by H. C. Ohanian and R. Ruffini: cf. OHANIAN – RUFFINI p. 299.

$$- \frac{1}{1 + \dfrac{2GM}{r} + \dfrac{\Lambda r^2}{3}} \, dr^2 +$$

$$- r^2 d\theta^2 - r^2 \sin^2 \theta \, d\varphi^2 \qquad (106)$$

where M is (a constant, that is equal to) the total mass of the system.

Therefore, in the case with $\Lambda > 0$ we have $\dfrac{2GM}{r} + \dfrac{\Lambda r^2}{3}$ instead of $\dfrac{2GM}{r}$, in the correct formulas.

Also in this case, as in that of the correct Schwarzschild solution for $\Lambda = 0$, we have not any singularity in the correct expression (106) for any value of $r > 0$. Therefore we can say that there is not any event horizon and therefore that there is not any black hole[72].

Furthermore, also here the redshift is always not infinite for any value of $r > 0$, as in the correct Schwarzschild solution for $\Lambda = 0$.

In particular, also here the coefficient of dt^2 is always ≥ 1, therefore since the coefficient of dt^2 is always not negative the time (that is, the temporal coordinate) does not become in any case as a spatial coordinate, contrary to the common treatment of the space-time inside the event horizon[73]. Moreover, since the coefficient of dr^2 is always not positive, also here the spatial coordinate r does not become in any case as a temporal

---

[72] Cf. OHANIAN – RUFFINI pp. 324-360; MISNER – THORNE – WHEELER pp. 819-940.

[73] Cf. OHANIAN – RUFFINI pp. 324-360; MISNER – THORNE – WHEELER pp. 819-940.

coordinate, contrary to the common treatment of the space-time inside the event horizon[74].

Therefore, also here the light cones are always orientated in the usual way, in particular there is not any horizontal inclination of the light cones, contrary to the common treatment of the space-time inside the event horizon[75].

The formulas (59), (60) and (71) in this case become:

$$\frac{d\tau_2}{d\tau_1} = \sqrt{\frac{1 + \dfrac{2GM}{r_1} + \dfrac{\Lambda r_1^2}{3}}{1 + \dfrac{2GM}{r_2} + \dfrac{\Lambda r_2^2}{3}}} \tag{107}$$

$$\frac{v_1}{v_2} = \sqrt{\frac{1 + \dfrac{2GM}{r_1} + \dfrac{\Lambda r_1^2}{3}}{1 + \dfrac{2GM}{r_2} + \dfrac{\Lambda r_2^2}{3}}} \tag{108}$$

$$v = \frac{dr}{dt} = 1 + \frac{2GM}{r} + \frac{\Lambda r^2}{3} \tag{109}$$

Therefore also in this case we have that in the presence of a gravitational field the clocks go more slowly (that is, that the measurements of time relatively to a reference frame that is integral with a space-time curved for the presence of a gravitational field flow more slowly). And the more intense is the gravitational field the more slowly the clocks go (that is, the

---

[74] Cf. OHANIAN – RUFFINI pp. 324-360; MISNER – THORNE – WHEELER pp. 819-940.

[75] Cf. OHANIAN – RUFFINI pp. 324-360; MISNER – THORNE – WHEELER pp. 819-940.

measurements of time relatively to a reference frame that is integral with a space-time curved for the presence of a more intense gravitational field flow more slowly than the measurements of time relatively to a reference frame that is integral with a space-time curved for the presence of a less intense gravitational field)[76].

Moreover, also here the fact that in the formula (109) v is > 1 (in cgs units, v is $> c$) produces a time beforehand of light. On the other hand, also here the fact that the velocity of light is > 1 does not violate the Special Theory of Relativity[77] because in this Theory the velocity of light is always 1 only in vacuum and relatively to inertial reference frames, and in this case we are not measuring in vacuum and relatively to an inertial reference frame, for the presence of the gravitational field (and for the fact that the cosmological constant is greater than zero) [that is, in this case there is not a vacuum for the presence of the gravitational field (and for the fact that the cosmological constant is greater than zero) and, moreover, we are not measuring

---

[76] Also here, of course, it is the clock positioned in a gravitational field to go more slowly than another clock positioned in a place without gravitational fields, not time itself: more precisely, we see the clocks to go more slowly in the presence of gravitational fields, because in these cases we use a reference frame that is integral with a space-time curved for the presence of the gravitational fields. On the other hand, it is physically wrong to use the measurements of a clock for measuring time without considering if the clock is in a gravitational field or not, and the intensity of the gravitational field if it is present, that is without considering if we are using a reference frame that is integral with a flat space-time or a reference frame that is integral with a space-time curved for the presence of the gravitational field, and the value of this curvature of the space-time (that is integral with the used reference frame) if it (this curvature) is present. Indeed, for effecting physical measurements of time we must use only well calibrated clocks and the calibration of the clocks obviously depends on the presence or the absence of a gravitational field, and on the intensity of the gravitational field if it is present, that is depends on the presence or the absence of a curvature in the space-time that is integral with the used reference frame, and on the value of this curvature if it is present (cf. PACE pp. 28-29).

[77] Cf. EINSTEIN (1905).

relatively to a reference frame that is integral with a flat space-time but we are measuring relatively to a reference frame that is integral with a space-time curved for the presence of the gravitational field (and for the fact that the cosmological constant is greater than zero)].

On the other hand, also here if we proceed by means of a proceeding that is the inverse of that with which we have transformed the expression (43) in the expression (48) or in the expression (49), for having the space-time interval expressed as a function of the coordinates t_i and r_i of an inertial reference frame, we obtain that in the expression of the space-time interval the coefficients of dt_i^2 and dr_i^2 are the inverses respectively of those of dt^2 and dr^2 in the expression (106), for which the radial velocity of light v_i relatively to an inertial reference frame results equal to:

$$v_i \; = \; \frac{dr_i}{dt_i} \; = \; \frac{1}{1 + \dfrac{2GM}{r} + \dfrac{\Lambda r^2}{3}} \tag{110}$$

that is to the inverse of the value in the non-inertial reference frame. Therefore, correctly, the value of v_i is always ≤ 1. Of course, in the formula (110) r is still relative to the non-inertial reference frame, but we can think of replacing [in the formula (110)] r point by point with the corresponding value of the inertial reference frame. However we have that, since $v_i = 1/v$, v_i can be ≤ 1, as the Special Theory of Relativity requires[78], only if $v \geq 1$. Therefore the formula (110) is in accord with the Special Theory of Relativity exactly because the formula (109) affirms that $v \geq 1$.

---

[78] Cf. EINSTEIN (1905).

2 The correct Kerr solution in the case in which the cosmological constant is greater than zero

Analogously, we can say that the correct Kerr solution in the case in which the cosmological constant Λ is greater than zero is expressed by this formula:

$$ds^2 = dt^2 - \frac{\rho^2}{\Delta} dr^2 - \rho^2 d\theta^2 +$$

$$- (r^2 + a^2) \sin^2 \theta \, d\varphi^2 +$$

$$+ \frac{2GMr + \frac{\Lambda}{3}r^4}{\rho^2} \{dt - a \sin^2 \theta \, d\varphi\}^2 \qquad (111)$$

where ρ^2 and Δ are functions of r and θ:

$$\rho^2 \equiv r^2 + a^2 \cos^2 \theta \qquad (112)$$

$$\Delta \equiv r^2 + 2GMr + \frac{\Lambda}{3}r^4 + a^2 \qquad (113)$$

and M and a are constants: M is the total mass of the system and a is the spin angular momentum of the system per unit mass.

Also in this case, as in that of the correct Kerr solution for $\Lambda = 0$, we have not any singularity in the correct expression (111) for any value of $r > 0$. Therefore we can say that there is not any event horizon and therefore that there is not any black hole[79].

---

[79] Cf. OHANIAN – RUFFINI pp. 324-360; MISNER – THORNE – WHEELER pp. 819-940.

Furthermore, also here the redshift is always not infinite for any value of $r > 0$, as in the correct Kerr solution for $\Lambda = 0$.

In particular, also here the coefficient of dt^2 is always ≥ 1, therefore since the coefficient of dt^2 is always not negative the time (that is, the temporal coordinate) does not become in any case as a spatial coordinate, contrary to the common treatment of the space-time inside the event horizon[80]. Moreover, since the coefficient of dr^2 is always not positive, also here the spatial coordinate r does not become in any case as a temporal coordinate, contrary to the common treatment of the space-time inside the event horizon[81].

Therefore also here the light cones are always orientated in the usual way, in particular there is not any horizontal inclination of the light cones, contrary to the common treatment of the space-time inside the event horizon[82].

In this case the formulas (81), (82) and (83) for $\sin^2 \theta = 0$ (and therefore for $\cos^2 \theta = 1$) become respectively:

$$\frac{d\tau_2}{d\tau_1} = \sqrt{\frac{1 + \dfrac{2GMr_1 + \frac{\Lambda}{3} r_1^4}{r_1^2 + a^2}}{1 + \dfrac{2GMr_2 + \frac{\Lambda}{3} r_2^4}{r_2^2 + a^2}}} \qquad (114)$$

---

[80] Cf. OHANIAN – RUFFINI pp. 324-360; MISNER – THORNE – WHEELER pp. 819-940.

[81] Cf. OHANIAN – RUFFINI pp. 324-360; MISNER – THORNE – WHEELER pp. 819-940.

[82] Cf. OHANIAN – RUFFINI pp. 324-360; MISNER – THORNE – WHEELER pp. 819-940.

$$\frac{v_1}{v_2} = \sqrt{\frac{1 + \dfrac{2GMr_1 + \frac{\Lambda}{3}r_1^4}{r_1^2 + a^2}}{1 + \dfrac{2GMr_2 + \frac{\Lambda}{3}r_2^4}{r_2^2 + a^2}}} \qquad (115)$$

$$v = \frac{dr}{dt} = 1 + \frac{2GMr + \frac{\Lambda}{3}r^4}{r^2 + a^2} \qquad (116)$$

Therefore also in this case we have that in the presence of a gravitational field the clocks go more slowly (that is, that the measurements of time relatively to a reference frame that is integral with a space-time curved for the presence of a gravitational field flow more slowly). And the more intense is the gravitational field the more slowly the clocks go (that is, the measurements of time relatively to a reference frame that is integral with a space-time curved for the presence of a more intense gravitational field flow more slowly than the measurements of time relatively to a reference frame that is integral with a space-time curved for the presence of a less intense gravitational field)[83].

---

[83] Also here, of course, it is the clock positioned in a gravitational field to go more slowly than another clock positioned in a place without gravitational fields, not time itself: more precisely, we see the clocks to go more slowly in the presence of gravitational fields, because in these cases we use a reference frame that is integral with a space-time curved for the presence of the gravitational fields. On the other hand, it is physically wrong to use the measurements of a clock for measuring time without considering if the clock is in a gravitational field or not, and the intensity of the gravitational field if it is present, that is without considering if we are using a reference frame that is integral with a flat space-time or a reference frame that is integral with a space-time curved for the presence of the gravitational field, and the value of this curvature of the space-time (that is integral with the used reference frame) if it (this curvature) is present. Indeed, for effecting physical measurements of time we must use only well calibrated clocks and the calibration of the clocks obviously depends on the presence or the absence of a gravitational field, and on

Moreover, also here the fact that in the formula (116) v is $>$ 1 (in cgs units, v is $> c$) produces a time beforehand of light. On the other hand, also here the fact that the velocity of light is > 1 does not violate the Special Theory of Relativity[84] because in this Theory the velocity of light is always 1 only in vacuum and relatively to inertial reference frames, and in this case we are not measuring in vacuum and relatively to an inertial reference frame, for the presence of the gravitational field (and for the fact that the cosmological constant is greater than zero) [that is, in this case there is not a vacuum for the presence of the gravitational field (and for the fact that the cosmological constant is greater than zero) and, moreover, we are not measuring relatively to a reference frame that is integral with a flat space-time but we are measuring relatively to a reference frame that is integral with a space-time curved for the presence of the gravitational field (and for the fact that the cosmological constant is greater than zero)].

On the other hand, also here if we proceed by means of a proceeding that is the inverse of that with which we have transformed the expression (43) in the expression (48) or in the expression (49), for having the space-time interval expressed as a function of the coordinates t_i and r_i of an inertial reference frame, we obtain that in the expression of the space-time interval the coefficients of dt_i^2 and dr_i^2 are the inverses respectively of those of dt^2 and dr^2 in the expression (111), for which the radial velocity of light v_i relatively to an inertial reference frame results equal to:

---

the intensity of the gravitational field if it is present, that is depends on the presence or the absence of a curvature in the space-time that is integral with the used reference frame, and on the value of this curvature if it is present (cf. PACE pp. 28-29).

[84] Cf. EINSTEIN (1905).

$$v_i = \frac{dr_i}{dt_i} = \frac{1}{1 + \dfrac{2GMr + \frac{\Lambda}{3}r^4}{r^2 + a^2}} \tag{117}$$

that is to the inverse of the value in the non-inertial reference frame. Therefore, correctly, the value of v_i is always ≤ 1. Of course, in the formula (117) r is still relative to the non-inertial reference frame, but we can think of replacing [in the formula (117)] r point by point with the corresponding value of the inertial reference frame. However we have that, since $v_i = 1/v$, v_i can be ≤ 1, as the Special Theory of Relativity requires[85], only if $v \geq 1$. Therefore the formula (117) is in accord with the Special Theory of Relativity exactly because the formula (116) affirms that $v \geq 1$.

3 The correct Reissner-Nordstrøm solution in the case in which the cosmological constant is greater than zero

Analogously, we can say that the correct Reissner-Nordstrøm solution in the case in which the cosmological constant Λ is greater than zero is expressed by this formula:

$$ds^2 = \left(1 + \frac{2GM}{r} + \frac{\Lambda r^2}{3} + \frac{GQ^2}{r^2}\right)dt^2 +$$

$$- \frac{1}{1 + \dfrac{2GM}{r} + \dfrac{\Lambda r^2}{3} + \dfrac{GQ^2}{r^2}}\,dr^2 +$$

---

[85] Cf. EINSTEIN (1905).

$$- r^2 d\theta^2 - r^2 \sin^2 \theta \, d\varphi^2 \qquad\qquad (118)$$

where M and Q are two constants: M is the total mass of the system and Q is the total electric charge of the system.

Also in this case, as in that of the correct Reissner-Nordstrøm solution for $\Lambda = 0$, we have not any singularity in the correct expression (118) for any value of $r > 0$. Therefore we can say that there is not any event horizon and therefore that there is not any black hole[86].

Furthermore, also here the redshift is always not infinite for any value of $r > 0$, as in the correct Reissner-Nordstrøm solution with $\Lambda = 0$.

In particular, also here the coefficient of dt^2 is always ≥ 1, therefore since the coefficient of dt^2 is always not negative the time (that is, the temporal coordinate) does not become in any case as a spatial coordinate, contrary to the common treatment of the space-time inside the event horizon[87]. Moreover, since the coefficient of dr^2 is always not positive, also here the spatial coordinate r does not become in any case as a temporal coordinate, contrary to the common treatment of the space-time inside the event horizon[88].

Therefore, also here the light cones are always orientated in the usual way, in particular there is not any horizontal inclination of the light cones, contrary to the common treatment of the space-time inside the event horizon[89].

In this case the formulas (88), (89) and (90) become respectively:

---

[86] Cf. OHANIAN – RUFFINI pp. 324-360; MISNER – THORNE – WHEELER pp. 819-940.

[87] Cf. OHANIAN – RUFFINI pp. 324-360; MISNER – THORNE – WHEELER pp. 819-940.

[88] Cf. OHANIAN – RUFFINI pp. 324-360; MISNER – THORNE – WHEELER pp. 819-940.

[89] Cf. OHANIAN – RUFFINI pp. 324-360; MISNER – THORNE – WHEELER pp. 819-940.

$$\frac{d\tau_2}{d\tau_1} = \sqrt{\frac{1 + \dfrac{2GM}{r_1} + \dfrac{\Lambda r_1^2}{3} + \dfrac{GQ^2}{r_1^2}}{1 + \dfrac{2GM}{r_2} + \dfrac{\Lambda r_2^2}{3} + \dfrac{GQ^2}{r_2^2}}} \qquad (119)$$

$$\frac{v_1}{v_2} = \sqrt{\frac{1 + \dfrac{2GM}{r_1} + \dfrac{\Lambda r_1^2}{3} + \dfrac{GQ^2}{r_1^2}}{1 + \dfrac{2GM}{r_2} + \dfrac{\Lambda r_2^2}{3} + \dfrac{GQ^2}{r_2^2}}} \qquad (120)$$

$$v = \frac{dr}{dt} = 1 + \frac{2GM}{r} + \frac{\Lambda r^2}{3} + \frac{GQ^2}{r^2} \qquad (121)$$

Therefore also in this case we have that in the presence of a gravitational field the clocks go more slowly (that is, that the measurements of time relatively to a reference frame that is integral with a space-time curved for the presence of a gravitational field flow more slowly). And the more intense is the gravitational field the more slowly the clocks go (that is, the measurements of time relatively to a reference frame that is integral with a space-time curved for the presence of a more intense gravitational field flow more slowly than the measurements of time relatively to a reference frame that is integral with a space-time curved for the presence of a less intense gravitational field)[90].

---

[90] Also here, of course, it is the clock positioned in a gravitational field to go more slowly than another clock positioned in a place without gravitational fields, not time itself: more precisely, we see the clocks to go more slowly in the presence of gravitational fields, because in these cases we use a reference frame that is integral with a space-time curved for the presence of the gravitational fields. On the other hand, it is physically wrong to use the measurements of a clock for measuring time

Moreover, also here the fact that in the formula (121) v is $>$ 1 (in cgs units, v is $> c$) produces a time beforehand of light. On the other hand, also here the fact that the velocity of light is > 1 does not violate the Special Theory of Relativity[91] because in this Theory the velocity of light is always 1 only in vacuum and relatively to inertial reference frames, and in this case we are not measuring in vacuum and relatively to an inertial reference frame, for the presence of the gravitational field (and also for the presence of the electromagnetic field and for the fact that the cosmological constant is greater than zero) [that is, in this case there is not a vacuum for the presence of the gravitational field (and also for the presence of the electromagnetic field and for the fact that the cosmological constant is greater than zero) and, moreover, we are not measuring relatively to a reference frame that is integral with a flat space-time but we are measuring relatively to a reference frame that is integral with a space-time curved for the presence of the gravitational field (and also for the presence of the electromagnetic field and for the fact that the cosmological constant is greater than zero)].

On the other hand, also here if we proceed by means of a proceeding that is the inverse of that with which we have transformed the expression (43) in the expression (48) or in the expression (49), for having the space-time interval expressed as a

---

without considering if the clock is in a gravitational field or not, and the intensity of the gravitational field if it is present, that is without considering if we are using a reference frame that is integral with a flat space-time or a reference frame that is integral with a space-time curved for the presence of the gravitational field, and the value of this curvature of the space-time (that is integral with the used reference frame) if it (this curvature) is present. Indeed, for effecting physical measurements of time we must use only well calibrated clocks and the calibration of the clocks obviously depends on the presence or the absence of a gravitational field, and on the intensity of the gravitational field if it is present, that is depends on the presence or the absence of a curvature in the space-time that is integral with the used reference frame, and on the value of this curvature if it is present (cf. PACE pp. 28-29).

[91] Cf. EINSTEIN (1905).

function of the coordinates t_i and r_i of an inertial reference frame, we obtain that in the expression of the space-time interval the coefficients of dt_i^2 and dr_i^2 are the inverses respectively of those of dt^2 and dr^2 in the expression (118), for which the radial velocity of light v_i relatively to an inertial reference frame results equal to:

$$v_i \;=\; \frac{dr_i}{dt_i} \;=\; \frac{1}{1+\dfrac{2GM}{r}+\dfrac{\Lambda r^2}{3}+\dfrac{GQ^2}{r^2}} \tag{122}$$

that is to the inverse of the value in the non-inertial reference frame. Therefore, correctly, the value of v_i is always ≤ 1. Of course, in the formula (122) r is still relative to the non-inertial reference frame, but we can think of replacing [in the formula (122)] r point by point with the corresponding value of the inertial reference frame. However we have that, since $v_i = 1/v$, v_i can be ≤ 1, as the Special Theory of Relativity requires[92], only if $v \geq 1$. Therefore the formula (122) is in accord with the Special Theory of Relativity exactly because the formula (121) affirms that $v \geq 1$.

4 The correct Kerr-Newman solution in the case in which the cosmological constant is greater than zero

Analogously, we can say that the correct Kerr-Newman solution in the case in which the cosmological constant Λ is greater than zero is expressed by this formula:

---

[92] Cf. EINSTEIN (1905).

$$ds^2 = \frac{\Delta}{\rho^2}[dt - a\,\sin^2\theta\,d\varphi]^2 +$$

$$-\frac{\sin^2\theta}{\rho^2}[(r^2 + a^2)d\varphi - a dt]^2 +$$

$$-\frac{\rho^2}{\Delta}dr^2 - \rho^2 d\theta^2 \tag{123}$$

where ρ^2 and Δ are functions of r and θ:

$$\Delta \equiv r^2 + 2GMr + \frac{\Lambda}{3}r^4 + a^2 + GQ^2 \tag{124}$$

$$\rho^2 \equiv r^2 + a^2 \cos^2\theta \tag{125}$$

and M, Q and a are constants: M is the total mass of the system, Q is the total electric charge of the system, and a is the angular momentum of the system per unit mass.

Also in this case, as in that of the correct Kerr-Newman solution for $\Lambda = 0$, we have not any singularity in the correct expression (123) for any value of $r > 0$. Therefore we can say that also here there is not any event horizon and therefore that there is not any black hole[93].

Furthermore, the redshift is always not infinite for any value of $r > 0$, as in the correct Kerr-Newman solution for $\Lambda = 0$.

In particular, also here the coefficient of dt^2 is always ≥ 1, therefore since the coefficient of dt^2 is always not negative the time (that is, the temporal coordinate) does not become in any case as a spatial coordinate, contrary to the common treatment of

---

[93] Cf. OHANIAN – RUFFINI pp. 324-360; MISNER – THORNE – WHEELER pp. 819-940.

the space-time inside the event horizon[94]. Moreover, since the coefficient of dr^2 is always not positive, also here the spatial coordinate r does not become in any case as a temporal coordinate, contrary to the common treatment of the space-time inside the event horizon[95].

Therefore, also here the light cones are always orientated in the usual way, in particular there is not any horizontal inclination of the light cones, contrary to the common treatment of the space-time inside the event horizon[96].

In this case the formulas (98), (99) and (100) for $\sin^2\theta = 0$ (and therefore for $\cos^2\theta = 1$) become respectively:

$$\frac{d\tau_2}{d\tau_1} = \sqrt{\frac{1 + \dfrac{2GMr_1 + \frac{\Lambda}{3}r_1^4 + GQ^2}{r_1^2 + a^2}}{1 + \dfrac{2GMr_2 + \frac{\Lambda}{3}r_2^4 + GQ^2}{r_2^2 + a^2}}} \tag{126}$$

$$\frac{v_1}{v_2} = \sqrt{\frac{1 + \dfrac{2GMr_1 + \frac{\Lambda}{3}r_1^4 + GQ^2}{r_1^2 + a^2}}{1 + \dfrac{2GMr_2 + \frac{\Lambda}{3}r_2^4 + GQ^2}{r_2^2 + a^2}}} \tag{127}$$

$$v = \frac{dr}{dt} = 1 + \frac{2GMr + \frac{\Lambda}{3}r^4 + GQ^2}{r^2 + a^2} \tag{128}$$

---

[94] Cf. OHANIAN – RUFFINI pp. 324-360; MISNER – THORNE – WHEELER pp. 819-940.

[95] Cf. OHANIAN – RUFFINI pp. 324-360; MISNER – THORNE – WHEELER pp. 819-940.

[96] Cf. OHANIAN – RUFFINI pp. 324-360; MISNER – THORNE – WHEELER pp. 819-940.

Therefore also in this case we have that in the presence of a gravitational field the clocks go more slowly (that is, that the measurements of time relatively to a reference frame that is integral with a space-time curved for the presence of a gravitational field flow more slowly). And the more intense is the gravitational field the more slowly the clocks go (that is, the measurements of time relatively to a reference frame that is integral with a space-time curved for the presence of a more intense gravitational field flow more slowly than the measurements of time relatively to a reference frame that is integral with a space-time curved for the presence of a less intense gravitational field)[97].

Moreover, also here the fact that in the formula (128) v is $>$ 1 (in cgs units, v is $> c$) produces a time beforehand of light. On the other hand, also here the fact that the velocity of light is > 1 does not violate the Special Theory of Relativity[98] because in this Theory the velocity of light is always 1 only in vacuum and relatively to inertial reference frames, and in this case we are not

---

[97] Also here, of course, it is the clock positioned in a gravitational field to go more slowly than another clock positioned in a place without gravitational fields, not time itself: more precisely, we see the clocks to go more slowly in the presence of gravitational fields, because in these cases we use a reference frame that is integral with a space-time curved for the presence of the gravitational fields. On the other hand, it is physically wrong to use the measurements of a clock for measuring time without considering if the clock is in a gravitational field or not, and the intensity of the gravitational field if it is present, that is without considering if we are using a reference frame that is integral with a flat space-time or a reference frame that is integral with a space-time curved for the presence of the gravitational field, and the value of this curvature of the space-time (that is integral with the used reference frame) if it (this curvature) is present. Indeed, for effecting physical measurements of time we must use only well calibrated clocks and the calibration of the clocks obviously depends on the presence or the absence of a gravitational field, and on the intensity of the gravitational field if it is present, that is depends on the presence or the absence of a curvature in the space-time that is integral with the used reference frame, and on the value of this curvature if it is present (cf. PACE pp. 28-29).

[98] Cf. EINSTEIN (1905).

measuring in vacuum and relatively to an inertial reference frame, for the presence of the gravitational field (and also for the presence of the electromagnetic field and for the fact that the cosmological constant is greater than zero) [that is, in this case there is not a vacuum for the presence of the gravitational field (and also for the presence of the electromagnetic field and for the fact that the cosmological constant is greater than zero) and, moreover, we are not measuring relatively to a reference frame that is integral with a flat space-time but we are measuring relatively to a reference frame that is integral with a space-time curved for the presence of the gravitational field (and also for the presence of the electromagnetic field and for the fact that the cosmological constant is greater than zero)].

On the other hand, also here if we proceed by means of a proceeding that is the inverse of that with which we have transformed the expression (43) in the expression (48) or in the expression (49), for having the space-time interval expressed as a function of the coordinates t_i and r_i of an inertial reference frame, we obtain that in the expression of the space-time interval the coefficients of dt_i^2 and dr_i^2 are the inverses respectively of those of dt^2 and dr^2 in the expression (123), for which the radial velocity of light v_i relatively to an inertial reference frame results equal to:

$$v_i \; = \; \frac{dr_i}{dt_i} \; = \; \frac{1}{1 + \dfrac{2GMr + \frac{\Lambda}{3}r^4 + GQ^2}{r^2 + a^2}} \qquad (129)$$

that is to the inverse of the value in the non-inertial reference frame. Therefore, correctly, the value of v_i is always ≤ 1. Of course, in the formula (129) r is still relative to the non-inertial reference frame, but we can think of replacing [in the formula (129)] r point by point with the corresponding value of the inertial reference frame. However we have that, since $v_i = 1/v$, v_i can be ≤ 1, as the Special Theory of Relativity requires[99], only if $v \geq 1$. Therefore the formula (129) is in accord with the Special Theory of Relativity exactly because the formula (128) affirms that $v \geq 1$.

---

[99] Cf. EINSTEIN (1905).

CHAPTER IV
THE VALIDITY OF THE GENERAL THEORY OF RELATIVITY IMPLIES IN GENERAL THE NON-EXISTENCE OF ANY BLACK HOLE

1 Also the Einstein's field equation of the General Theory of Relativity implies in general the non-existence of any event horizon and therefore the non-existence of any black hole

The Einstein's field equation of the General Theory of Relativity[100] is symmetric with respect to time[101], therefore we can always reverse the motion, that is due to the gravitational field, of any mass in such a way as if it went back in time simply by means of changing the sign of the velocity of that mass, that is by means of reversing the direction of the motion of that mass. (If also other masses move for the gravitational forces we would have to change the sign of the velocities of all these masses for reversing the global motion of all these masses, in such a way as if they went back in time: this would be only a more complicated case but the result would be always the same). This implies in

---

[100] Cf. EINSTEIN (1916); OHANIAN – RUFFINI pp. 275-293; MISNER – THORNE – WHEELER pp. 383-590; WEINBERG (1972) pp. 151-171; ANDERSON pp. 329-371.

[101] Cf. HAWKING (1997) p. 97.167.176; DAVIES (1997) pp. 229-231; BARROW (2003) pp. 284-286; BARROW (1996) pp. 167-168; PENROSE (2009) pp. 386-389; GOTT p. 197.

general the non-existence of any event horizon, since the existence of an event horizon would imply the impossibility of going back for any mass once this mass has gone from the outside of the event horizon in the inside of the same event horizon and we have seen that the Einstein's field equation (of the General Theory of Relativity) implies always for any mass the possibility of going back by means of reversing the direction of the motion of the same mass. Therefore, according to the Einstein's field equation (of the General Theory of Relativity) the event horizons, and consequently also the black holes, cannot exist in any case.

In other words, the non-existence of any event horizon defends the symmetry of the Einstein's field equation (of the General Theory of Relativity) with respect to time. On the other hand, the existence of an event horizon due to a gravitational field would demonstrate the incorrectness of the Einstein's field equation (of the General Theory of Relativity). Therefore, if we consider valid the General Theory of Relativity we must consider completely impossible the existence of any event horizon due to gravitational fields and consequently we must consider completely impossible the existence of any black hole due to gravitational fields.

Moreover, on the basis of the General Theory of Relativity we cannot justify the existence of the arrow of time by means of the existence, or of the possibility of existence, of event horizons.

2 The motion of a little mass in the case of the correct Schwarzschild solution (with $\Lambda = 0$)

Now, we examine the dynamics of a little mass P in the case of the correct Schwarzschild solution (when $\Lambda = 0$). For simplicity we will consider negligible the possible irradiation from the same little mass P because of its acceleration. A motionless little mass P, that has a rest mass $m \ll M$, at $r = +\infty$ has a gravitational potential energy equal to zero. If this little mass P moves because of the effect of the gravitational field in the direction of the central mass M, the same little mass P acquires a kinetic energy equal to the decrease of its gravitational potential energy that becomes negative. Therefore, during the motion of P, the total energy of P remains always the same and consequently both the inertial mass of P and the gravitational mass of P remain always the same. For $r \to 0$ we have that the kinetic energy of P tends to $+\infty$ according to the relativistic formulas (of course, without exceeding the velocity of the light in vacuum c relatively to any inertial reference frame) while the gravitational potential energy of P tends to $-\infty$, in such a way as to maintain constant the total energy of P and therefore (to maintain constant) also both the gravitational mass of P and the inertial mass of P. Obviously, in any moment of this motion of P, if we reverse the direction of the motion of P, P would return to its initial motionless state at $r = +\infty$.

In real situations the little mass P can irradiate and can interact with other particles, and therefore can decrease or can increase in its total energy: but, in any case, its total energy cannot decrease to negative values and if its total energy becomes zero we have in reality that P by means of its interactions has been transformed in something else.

On the other hand, if we consider a photon, instead of a little mass P with rest mass that is different from zero, in the case that we suppose that the photons have a very little rest mass that is compatible with the experimental results, then we can proceed in the same way. On the contrary, in the case that we suppose that the rest mass of the photons is rigorously zero, then the total energy of the photon at $r = +\infty$ corresponds to the rest mass of the little mass P. Moreover the photon instead of increasing its velocity increases its frequency during its motion towards the central mass M in such a way that its inertial mass and its gravitational mass remain always the same during the motion. Analogously, as for the possible return back of the photon to $r = +\infty$, the photon, during the return back, will decrease its frequency instead of decreasing its velocity, always in such a way that the its inertial mass and its gravitational mass remain always the same during the motion. On the other hand, this reasoning is obviously valid not only for the photons but also in an analogous way for the other particles that are considered to have rest mass equal to zero.

Furthermore, obviously, in the cases (with $\Lambda = 0$) of the correct Kerr solution, of the correct Reissner-Nordstrøm solution and of the correct Kerr-Newman solution, the little masses and the particles (with or without rest mass) behave in an analogous way to the way in which they behave in the case (with $\Lambda = 0$) of the correct Schwarzschild solution.

CHAPTER V
SOME OTHER IMPORTANT PHYSICAL CONSEQUENCES OF MY CORRECTIONS

1 Consequences on entropy and on the Hawking emission process

As for entropy, since it is not possible any event horizon we must always apply the traditional formulas for the calculation of entropy. Moreover, we can say that the formulas of Bekenstein and Hawking for the value of the entropy of black holes[102] regard objects that are impossible according to the General Theory of Relativity, since we have seen that the black holes, as have been defined in physics, are incompatible with the General Theory of Relativity.

In any case the symmetry of the Einstein's field equation (of the General Theory of Relativity) with respect to time guarantees that any physical process due to the General Theory of Relativity (since any process of this type, as we have seen, is theoretically reversible) conserves always the information on the physical system, contrary to some contemporary theories about black holes[103].

Analogously, as for thermodynamics, since it is not possible any event horizon, we must always apply the traditional formulas of thermodynamics, in particular the Planck's law of black-body radiation. Moreover, also here we can say that the Hawking

---

[102] Cf. OHANIAN – RUFFINI pp. 360-367.
[103] Cf. HAWKING – PENROSE pp. 49.53.74.79-81.151.157.

emission process for the evaporation of black holes[104] regards objects that are impossible according to the General Theory of Relativity, since we have seen that the black holes, as have been defined in physics, are incompatible with the General Theory of Relativity.

2 The impossibility of any event horizon and the Big Bang Theory

The impossibility of any event horizon according to the General Theory of Relativity also permits to explain in a simple way the possibility of the Big Bang[105].

In fact, according to the Big Bang Theory the total energy of the Universe was concentrated in an infinitesimal space and therefore the primordial Universe, according to the common point of view on the General Theory of Relativity, was substantially a supermassive black hole, even though commonly the relativistic physicists refuse to use the name of black hole for the entire Universe, since they know that according to the Big Bang Theory the primordial Universe has not behaved as a black hole.

Now, if the primordial Universe was a black hole not only we would have had the presence of an event horizon surrounding the primordial Universe, contrary to the Big Bang Theory, but also we would have had that, according to the current theory

---

[104] Cf. OHANIAN – RUFFINI pp. 360-367; HAWKING – PENROSE pp. 49-74; HAWKING (1997) pp. 120-135.

[105] Cf. OHANIAN – RUFFINI pp. 389-437.444-473; MISNER – THORNE – WHEELER pp. 703-799; WEINBERG (1972) pp. 469-597; ANDERSON pp. 444-465; HAWKING – PENROSE pp. 30.37-47.91-122; BARROW (1998) pp. 11-125; DAVIES (2000) pp. 30-46; HAWKING (1997) pp. 51-70.136-165; WEINBERG (1980) pp. 14-165; DAVIES (1997) pp. 133-177.200-214; GOTT pp. 153-197.

about black holes[106], the primordial Universe would have not expanded in any case: in fact, according to the common theory about black holes[107], inside the event horizon of every black hole all the light cones are orientated towards the centre of the same black hole and therefore there is not any possibility of an expansion of the central mass of a black hole, while the Big Bang Theory affirms that there was a very strong expansion of the primordial Universe.

In conclusion, my demonstration, by starting from the validity of the General Theory of Relativity, of the impossibility of the existence of an event horizon due to a gravitational field permits of affirming the validity of the Big Bang Theory without going against the known laws of physics and in particular without going against the General Theory of Relativity.

Analogously, my correction to the commonly accepted theory permits a cyclic Universe[108] with a Big Bang and then a Big Crunch[109] and then a new Big Bang and then a new Big Crunch and so on: in fact, also a Big Crunch is not a black hole and is not surrounded by an event horizon, therefore every Big Crunch is theoretically reversible. On the other hand, the entropy of the Universe would continue to increase for every cycle Big Bang-Big Crunch[110] and therefore the existing Universe, that has a very small entropy, for thermodynamic reasons should have had origin in a finite past, also in the hypothesis that the existing Universe be a cyclic Universe.

---

[106] Cf. OHANIAN – RUFFINI pp. 324-360; MISNER – THORNE – WHEELER pp. 819-940.

[107] Cf. OHANIAN – RUFFINI pp. 324-360; MISNER – THORNE – WHEELER pp. 819-940.

[108] Cf. OHANIAN – RUFFINI pp. 418-428; BARROW (1998) pp. 37-39; DAVIES (2000) pp. 145-150.

[109] Cf. HAWKING – PENROSE pp. 37.44.106-107.117.125; BARROW (1998) pp. 18.37.72; HAWKING (1997) pp. 136.195; DAVIES (2000) pp. 124-131.145-150; WEINBERG (1980) pp. 166-170.

[110] Cf. OHANIAN – RUFFINI p. 422; BARROW (1998) pp. 31-39; DAVIES (2000) pp. 145-150.

CHAPTER VI
EXPERIMENTAL PROSPECTS

As we have seen, the experimental results at present are compatible both with my correction to the commonly accepted theory (that is, with my theory) and with the commonly accepted theory.

On the other hand, since my correction (to the commonly accepted theory) implies differences that can be measured by means of future experiments, we could test the predictions of my theory by means of future experiments that have the aim of testing my predictions. In particular, in the usual case of $\frac{2GM}{r} \ll 1$ we could measure the slowing down of clocks in gravitational fields (that is, the fact that the measurements of time relatively to a reference frame that is integral with a space-time curved for the presence of a more intense gravitational field flow more slowly than the measurements of time relatively to a reference frame that is integral with a space-time curved for the presence of a less intense gravitational field) at the second order in $\frac{2GM}{r}$ for testing my theory in comparison with the commonly accepted theory.

Moreover, according to my theory there is not any black hole due to gravitational fields, and therefore, according to my theory, certainly we will not see any black hole due to gravitational fields, contrary to the commonly accepted theory and to the common expectations of the present-day relativistic physicists. Obviously, if we will see a mass M in a sphere of radius smaller than $2GM$ (in cgs units, smaller than $\frac{2GM}{c^2}$) this will not imply automatically the existence of a black hole, since in my theory these concentrations of mass do not produce a black hole in any way. Instead, in this case we should measure the quantities that are different according to the two theories, for seeing which of these two (theories) is the right theory. However, in this case the difference between my theory and the commonly accepted theory is very clear: if that sphere did not contain a black hole, my theory would be correct and the commonly accepted theory would be incorrect. Vice versa if that sphere contained a black hole due to gravitational fields, then the commonly accepted theory would be confirmed and my theory would be incorrect.

In any case, I have founded my correction to the commonly accepted theory also on the basis of the Einstein's field equation of the General Theory of Relativity, and therefore the experimental tests that are in favour of the Einstein's field equation (of the General Theory of Relativity) are also in favour of my correction to the commonly accepted theory. And, since the experimental tests in favour of the Einstein's field equation (of the General Theory of Relativity) are generally considered more than enough for considering valid the Einstein's field equation (of the General Theory of Relativity), we can say that these same tests are more than enough also for considering valid my correction to the commonly accepted theory.

Finally, the fact that my correction to the commonly accepted theory is more compatible with the Big Bang Theory than the commonly accepted theory is another strong support in favour of my position, since the Big Bang Theory is generally considered to be a theory with strong experimental supports.

GENERAL CONCLUSION: ACCORDING TO THE GENERAL THEORY OF RELATIVITY, THERE IS NOT ANY EVENT HORIZON AND THEREFORE THERE IS NOT ANY BLACK HOLE

In conclusion, we have seen that the correct solutions of the Einstein's field equation (of the General Theory of Relativity) show, contrary to the corresponding incorrect solutions that are commonly used by the relativistic physicists, that in them (in the correct solutions) there is not any event horizon and therefore that in them there is not any black hole.

Moreover, we have seen that general considerations about the form of the Einstein's field equation (of the General Theory of Relativity) demonstrate that in general any solution of this equation cannot include an event horizon and therefore cannot include a black hole.

Therefore, on the basis of what we have seen, we can say that, according to the General Theory of Relativity, there is not any event horizon and therefore there is not any black hole. In other words, according to the General Theory of Relativity, there is not any event horizon due to a gravitational field and therefore there is not any black hole due to a gravitational field.

As we have seen, since my correction to the commonly accepted theory is based on the Einstein's field equation (of the General Theory of Relativity), I am not in need of experimental confirmations to my theory.

But, since my theory implies experimental results that are different from those that are implied by the commonly accepted theory, my theory can also be tested experimentally; and, since an experimental confirmation of my theory would eliminate any doubt about the validity of my theory, it would be a good thing to effectuate an experimental test for deciding between the commonly accepted theory and my theory, that is substantially a correction of the commonly accepted theory.

Finally, my theory implies that all the physics that is based on the existence, or on the possibility of existence, of event horizons and black holes is incorrect. Therefore, according to my theory, this physics should be completely rewritten on the basis of this book and in particular on the basis of the correct Schwarzschild solution, of the correct Kerr solution, of the correct Reissner-Nordstrøm solution and of the correct Kerr-Newman solution, both in the case of $\Lambda = 0$ and in the case of $\Lambda > 0$.

BIBLIOGRAPHY

ANDERSON J. L., *Principles of Relativity Physics*, Academic Press, New York 1967.

BARROW J. D., *Il mondo dentro il mondo*, Adelphi, Milano 1996[3] {Italian translation of BARROW J. D., *The World within the World*, Oxford University Press, Oxford 1988}.

BARROW J. D., *Le origini dell'Universo*, Sansoni, Milano 1998[3] {Italian translation of BARROW J. D., *The Origin of the Universe*, The Orion Publishing Group Ltd., London 1994}.

BARROW J. D., *Teorie del tutto: La ricerca della spiegazione ultima*, Adelphi, Milano 2003[3] {Italian translation of BARROW J. D., *Theories of Everything: The Quest for Ultimate Explanation*, Oxford University Press, Oxford 1991}.

DAVIES P., *The Edge of Infinity: Where the Universe Came from and how it Will End*, Simon & Schuster, New York 1981.

DAVIES P., *I misteri del tempo*, Oscar Saggi Mondadori, Milano 1997 {Italian translation of DAVIES P., *About Time*, Orion Productions, Adelaide 1995}.

DAVIES P., *Gli ultimi tre minuti*, Superbur Scienza, Milano 2000 {Italian translation of DAVIES P., *The Last Three Minutes*, The Orion Publishing Group Ltd., London 1994}.

EINSTEIN A., *Zur Elektrodynamik bewegter Körper*, in «Annalen der Physik» 17 (1905) 891-921 {English translation: EINSTEIN A., *On the Electrodynamics of Moving Bodies*, in LORENTZ H. A. – EINSTEIN A. – MINKOWSKI H. – WEYL H., *The Principle of Relativity*, Dover Publications, New York 1952, pp. 35-65}.

EINSTEIN A., *Die Grundlage der allgemeinen Relativitätstheorie*, in «Annalen der Physik» 49 (1916) 769-822 {English translation: EINSTEIN A., *The Foundation of the General*

Theory of Relativity, in LORENTZ H. A. – EINSTEIN A. – MINKOWSKI H. – WEYL H., *The Principle of Relativity*, Dover Publications, New York 1952, pp. 109-164}.

GOTT J. R. III, *Viaggiare nel tempo*, Mondadori, Milano 2002 {Italian translation of GOTT J. R. III, *Time Travel in Einstein's Universe*, Houghton Mifflin Harcourt, Boston 2002}.

HAWKING S., *Dal Big Bang ai Buchi Neri*, BUR Supersaggi, Milano 1997[15] {Italian translation of HAWKING S., *A Brief History of Time*, Bantam Dell Publishing Group, New York 1988}.

HAWKING S., *The Universe in a Nutshell*, Bantam Spectra, New York 2001.

HAWKING S. W. – PENROSE R., *La Natura dello Spazio e del Tempo*, Sansoni, Milano 1996[2] {Italian translation of HAWKING S. W. – PENROSE R., *The Nature of Space and Time*, Princeton University Press, Princeton 1996}.

MISNER C. W. – THORNE K. S. – WHEELER J. A., *Gravitation*, W. H. Freeman and Company, New York 1973.

OHANIAN H. C. – RUFFINI R., *Gravitation and Spacetime*, Cambridge University Press, New York 2013[3].

PACE C. M., *All the relativistic temporal paradoxes are completely false*, Youcanprint, Tricase (LE) 2016.

PENROSE R., *La mente nuova dell'imperatore*, BUR Scienza, 2009[5] {Italian translation of PENROSE R., *The Emperor's New Mind*, Oxford University Press, Oxford 1989}.

PENROSE R., *Cycles of Time: An Extraordinary New View of the Universe*, The Bodley Head, London 2010.

REGGE T., *Infinito: Viaggio ai limiti dell'universo*, Oscar Saggi Mondadori, Milano 1996.

SCHWARZSCHILD K., *Über das Gravitationsfeld eines Massenpunktes nach der EINSTEINschen Theorie*, in «Sitzungsberichte der Königlich-Preussischen Akademie der Wissenschaften» 7 (1916) 189-196.

Susskind L., *The Black Hole War*, Little, Brown and Company, New York 2008.

Thorne K. S., *Black Holes and Time Warps: Einstein's Outrageous Legacy*, W. W. Norton & Company, New York 1994.

Weinberg S., *Gravitation and Cosmology: Principles and Applications of the General Theory of Relativity*, John Wiley & Sons, New York 1972.

Weinberg S., *I primi tre minuti*, Oscar Saggi Mondadori, Milano 1980 {Italian translation of Weinberg S., *The First Three Minutes: A Modern View of the Origin of the Universe*, Basic Books, New York 1977}.

TABLE OF CONTENTS

Finito di stampare nel mese di Dicembre 2018
per conto di Youcanprint *Self-Publishing*